STUDENT'S SOLUTIONS MANUAL

David Lund

University of Wisconsin—Eau Claire

Statistical Reasoning For Everyday Life

SECOND EDITION

Jeffrey O. Bennett ■ William L. Briggs ■ Mario F. Triola

Boston San Francisco New York
London Toronto Sydney Tokyo Singapore Madrid
Mexico City Munich Paris Cape Town Hong Kong Montreal

Reproduced by Addison-Wesley from camera-ready copy supplied by the authors.

Printed in the United States of America.

ISBN 0-201-83846-X

1 2 3 4 5 6 7 8 9 10 PHC 05 04 03 02

PREFACE

Student's Solutions Manual to Accompany Statistical Reasoning for Everyday Life by Jeffrey Bennett, William Briggs, and Mario Triola provides solutions to every odd-numbered exercise in Chapters One through Ten of the text and to all of the Review Exercises at the end of each chapter. The solutions are more than just answers -- reasoning and intermediate steps in the process of solving the exercises are also presented.

I want to express my thanks to my wife, Judy, for her support and patience throughout this project, and to Anna Stillner and the rest of the staff at Addison Wesley who have been such a great help in bringing this work to completion.

CONTENTS

CHAPTER 1 ANSWERS

Section 1.1

1 This statement does not make sense. The sample consists of the people actually studied, and it is not possible for the IRS to study every single adult in the United States.

3 This statement is sensible. It suggests that Smith had a substantial lead two weeks before the election, but leads can certainly evaporate in two weeks.

5 This statement does not make sense. The population of interest is people who have suffered a family tragedy, not people who may have had a recent cold.

7 **Population:** All registered voters in California
Sample: The 1026 people who were interviewed
Raw data: The responses from 1026 people who were interviewed
Sample statistics: Measures of the opinions about candidates, such as the percentage of people who like or dislike each candidate or the percentage of people who would vote for each candidate, among the 1026 people in the sample
Population parameters: Measures of the opinions about candidates in the population of all registered voters in California

9 **Population:** All new computers of this particular model from Dell
Sample: The single computer tested
Raw data: The times recorded for the various tasks in the test
Sample statistics: Measures that describe the test results for the sample computer in terms of speed and other "benchmarks"
Population parameters: Measures that describe the speed and other "benchmarks" for all computers in the population

11 a) The range is 4.5 $\pm$ .4 or 4.1 to 4.9 months.
b) The range is 95% $\pm$ .1% or 94.9% to 95.1%.
c) The range is 457 $\pm$ 1.2 or 455.8 to 458.2.

13 With a sample statistic of 53% and a margin of error of 2.5 percentage points, we are 95% confident that the interval from 50.5% to 55.5% contains the population parameter which is the true percentage of the population of voters who prefer the Republican candidate. Thus the Republican candidate can be at least 95% certain of winning the election and it would be prudent to plan a victory party.

15 With a sample statistic of 70% and a margin of error of 3 percentage points, we are 95% confident that the interval from 67% to 73% contains the population parameter which is the true percentage of the voters who would say that they voted in the recent presidential election. This entire range is, however, somewhat higher than the actual 61% who voted according to the voting records. This suggests that there were some people in the sample who did not actually vote, but said that they did when polled. While it is still possible (as always) that this particular sample is unusual and everyone told the truth, the lower end

of the range (67%) is quite far from 61%, making this an unlikely possibility.

17 a) **Goal:** Determine whether people think that they must rely only on their own skills and abilities to get ahead
Population: All people in the U.S.
Population parameter: The percentage of people in the population who believe that they must rely on themselves

b) **Sample:** The 4000 adults surveyed
Raw data: The individual responses to the question
Sample statistic: The percentage of people in the sample who believe that they must rely on themselves

c) The range of values likely to contain the population parameter is 70% ± 1.6% or 68.4% to 71.6%.

19 a) **Goal:** Determine the mean height of male Marine recruits
Population: All male Marine recruits
Population parameter: The mean height of male Marine recruits

b) **Sample:** The 772 male recruits measured
Raw data: The individual heights of the recruits in the sample
Sample statistic: The mean height of the 772 male marine recruits

c) The range of values likely to contain the population parameter is 69.7inches ± 0.2inches or 69.5 to 69.9 inches.

21 a) **Goal:** Determine the percentage of adults in the U.S. who keep money in regular savings accounts
Population: All adults in the U.S.
Population parameter: The percentage of all adults in the U.S. who keep money in regular savings accounts

b) **Sample:** The 2000 adults surveyed
Raw data: The individual responses of the adults in the sample
Sample statistic: The percentage of adults in the sample who keep money in regular savings accounts

c) The range of values likely to contain the population parameter is 64% ± 2.0% or 62% to 66%.

23 Step 1: Goal: Predict the winner of the next election for class. president. Population: All of the students who are likely to vote.

Step 2: Choose a sample of students who are likely to vote.

Step 3: Interview the students in the sample to determine the candidate for whom they intend to vote. Obtain the percentage of those in the sample who intend to vote for each candidate and determine what the election outcome would be if it were based solely on your sample.

Step 4: Use statistical techniques to infer the likely results for the entire population of students in the class.

Step 5: Based on the likely population results, draw conclusions about who will win the election.

25 Step 1: Goal: Determine what is the typical percentage of a restaurant bill that is left as a tip. Population: All restaurant bills, perhaps in a particular city or at a particular restaurant.

Step 2: Choose a sample of bills from your population and record the amount left as a tip.

Step 3: Find the percentage tip on each bill in your sample. Then calculate a mean tip for all the bills.

Step 4: Use statistical techniques to infer the likely mean percentage tip for the population.

Step 5: Based on the likely mean tip, draw conclusions about the mean percentage tip for all bills in your population. [Note: This is a little more involved than it appears to be at first glance since the mean percentage tip will be the sum of all tips divided by the sum of all bills, not just the mean of the percentages.]

27 Step 1: Goal: Determine the percentage of high school students who are vegetarians. Population: All high school students, perhaps in particular city or school.

Step 2: Choose a sample of high school students from your population.

Step 3: Interview the students in your sample to determine whether they are vegetarians. Then calculate the percentage of students in your sample who are vegetarians.

Step 4: Use statistical techniques to infer the likely percentage of all high school students in your population who are vegetarians.

Step 5: Draw conclusions about how many high school students are vegetarians.

Section 1.2

1 This statement does not make sense. A census would mean studying every teenage babysitter in the country (or world), which is clearly not practical.

3 This statement makes sense. It's quite apparent that most Americans are not more than 6 feet tall, so a study that comes to a ridiculous conclusion must have suffered from some form of bias.

5 This statement does not make sense. Stratified sampling involves using strata that differ in some known characteristics, such as male/female or state of origin. The approach described in this question will yield groups chosen randomly, and thus does not represent stratified sampling.

7 A census is practical. You will need to find a way to get reliable data, such as viewing the transcripts of all 50 students or, because of privacy issues, having a school official provide the GPAs without the names.

9 A census would require determining the energy cost of every home in Missouri, which requires gathering data from every home's utility and other energy bills and wood heating costs. The data from utilities should be available in principle, but obtaining additional data from oil and gas companies and combining the data for each home in the state is not very practical.

11 The entire team is the best sample. The other groups are specialized subsets of the entire team, and their eating habits and/or training methods may be quite different from those of the team as a whole and thus not representative.

13 The critic may be under real or imagined pressure to give a favorable review to the film since she works for the same company that produced the film.

15 The Book Review section generates revenue from advertisements, and therefore has a vested financial interest in making advertisers happy. It is possible that this could translate into giving better reviews to an advertised book than the book would otherwise get.

17 The university scientists are receiving funding from Monsanto, which might make them eager to please Monsanto in hopes of getting additional funding opportunities in the future. Thus, there is a potential for bias toward giving Monsanto the results it wants, even though they do not work for Monsanto.

19 The production process may produce a number of chips at the same time, with the chips coming off the production line in the same order each time. For example, if 10 chips are made simultaneously and the seventh one is always defective, then sampling every 100th chip will either result in a defective chip every time (if the seventh one is always chosen) or it will result in missing the always defective chip (if the seventh one is not chosen). In the first case, the percentage of defective chips will be overestimated; in the latter case, it will be underestimated. In general, it is not a good idea to use systematic sampling on any product that is produced systematically. If the production process is not systematic, then sampling every 100th chip should yield a representative sample.

21 The poll involves only very early morning voters, who may not be representative of all voters since they may be voting early for a reason such as being employed or getting children off to school. Such a poll might easily exclude many retired or unemployed voters who are not in a hurry to vote.

23 Because the students can choose whether to respond by returning the surveys, those with strong opinions are more likely to respond. This survey is very unlikely to be representative of all students.

25 The employees of the company have a vested interest in the product's success and this bias make them a poor sample.

27 Customers who shop at a supermarket between 10AM and noon are more likely to be people who are not employed, so this sample is unlikely to be representative of all customers.

29 This is a convenience sample, and it is one that is unlikely to be representative because people with strong feelings are more likely to return the survey. The magazine probably chose this sampling method because it was easy; the magazine might even be interested in the opinions of those with the strongest feelings.

31 This is stratified sampling in which the strata are the different age groups. This method probably was chosen to see whether sleep habits vary with age.

33 This is cluster sampling in which the clusters are the homes on each of the three streets chosen. There is no indication that the streets were chosen at random, so perhaps they were chosen because they were deemed to be "typical" and likely to produce a representative sample.

35 This is stratified sampling in which the strata are the different states. This is necessary since the study is one of how incomes vary among the states.

37 Simple random sampling should be adequate. However, stratified sampling in which the strata are different ethnic groups is also a possibility. This would enable one to gather information about the differences in percentages of blood types among the different ethnic groups.

39 Simple random sampling of lung cancer victims should be adequate if the population of such victims can be identified. However, you might want to do cluster sampling from coroners' records in case there are regional differences in the causes of lung cancer.

41 You will need stratified sampling in which you study both people who drink three cups of herbal tea and people who do not. This study would be best done as a controlled experiment.

Section 1.3

1 This statement is sensible. Such a study would necessarily be observational.

3 This statement is not sensible. While the lab would use a procedure to test for strep, it would not be considered a statistical experiment.

5 This statement is not sensible. A meta-analysis involves combining several existing studies; a census is needed to determine the population of New Zealand.

7 This is an observational, case-control study. The cases were the identical twins and the controls were the fraternal twins.

9 This is an observational, case-control study. The cases were the melanoma patients and the controls were the cancer-free patients.

11 This is an observational study involving a poll.

13 This is an observational, case-control study. The cases are the women who exercise regularly and the controls are those women who did not.

15 This is a meta-analysis combining the results of 11 separate studies.

17 This is a meta-analysis combining the results of 72 separate studies.

19 This experiment should be effective. It might be advisable for the experiment to be blind so that the researchers evaluating the growth do not know which group received which food.

21 This experiment suffers because the two groups live in different states with different driving conditions and different weather. If it turns out that there is a difference in the condition of the cars in the two groups, it will not be possible to determine whether that result was due to the different waxes or to the different driving conditions. This is an example of confounding. The groups should have been randomized so that people from each state could end up in each group.

23 The idea behind this experiment is reasonable, but measurement difficulties introduce many opportunities for confounding. The amount of sleep is determined by what time people go to sleep as well as when they awake. Since people don't know exactly what time they will fall asleep, it will be difficult for them to set their alarms to ensure that they get 10% less sleep than they do usually. It may also be difficult to evaluate the number of temper tantrums of each subject.

25 This should be an experiment in which students are randomly assigned to treatment groups that listen to jazz music or other music or to a control group that does not. This study cannot be single- or double-blind.

27 Ideally, this should be an experiment in which swimmers are randomly assigned to a treatment group that plays soccer or a control group that does not. However, some swimmers may have no interest in playing soccer, so an observational case-control study may be necessary with the cases consisting of those who play soccer and the controls consisting of those who do not. In either case, it is possible that there will be other confounding effects due to other types of exercise that may be as beneficial or more beneficial than playing soccer, such as running or weight lifting.

29 This should be an experiment in which the mind-reader is asked to describe something that another person is thinking, repeated with many people. The experiment should be double-blind so that the mind reader does not know the identity of the person he is "reading" and the person, when asked what he or she is thinking, does not know what the mind-reader has said.

Section 1.4

1 Guideline #2 may be an issue. If one of the brands was a Wholesome Foods brand, then the public relations department has an interest in the outcome and should not be conducting the experiment. If no Wholesome Foods brands were involved in the taste test, then there is no conflict of interest.

3 Guidelines #4 and #6 are at issue here. Is the researcher interested in how the adults felt on most mornings or on the particular morning of the day they were contacted? Did the researcher try to determine whether illness was a factor in not feeling well? Is the researcher interested in physical health or mental health?

5 Guideline #6 is at issue here. The question suggests a possible response, rather than just asking respondents to name their favorite beer.

7 The two questions are different. For example, some people might think that abortion is wrong, but still favor choice. Such people would be likely to say "yes" to the first question while favoring choice in the second. The wording of the second question might also lead to confusion among some of those interviewed, as they may think "advice of her doctor" means the woman's life is in danger, which could alter their opinion of whether abortion is justified in the situation. Groups opposed to abortion would be likely to cite the results of the first question, while groups favoring choice would be likely to cite the results of the second question.

9 The headline says "drugs" whereas the story says "drug use, drinking, or smoking." Because "drugs" is usually taken to mean drugs *other than* smoking or alcohol, the headline is very misleading. There is also a difference between "98% of movies" and "98% of top movie rentals." The latter group is much smaller than the former.

11 The first question requires a study of blind dates generally, while the second examines people who are married to see whether their first date was blind. The first question is a more difficult one to study since, at the time of the study, some blind dates will not yet have led to marriage, but may eventually. In addition, there is no good way to determine who is in the population of blind daters.

13 The first question asks what percentage of the time do teenagers run red lights and would require observations of many situations in which teenagers encounter a red light. The second question asks what percentage of red light runners are teenagers and would require ascertaining the age of each of those observed to have run a red light. The percentage results are likely to be quite different.

15 The report seems to be making an implication of restaurant quality in New York (the "Big Apple"), but there is nothing unusual about the case of New York City. With only nine scores of 29, most large cities will not have a restaurant with a score of 29. In addition, data are missing. What about restaurants receiving scores of 30, or 28, or 27? What criteria were used for the ratings? Without much more information, it would be difficult to act on these data.

Chapter Review Exercises

1
- a) The likely range is 93.9% $\pm$ 1.3% or 92.6% to 95.2%.
- b) The population consists of all U.S. households.
- c) A survey is an observational study.
- d) The reported value is a sample statistic based on the 1123 households surveyed, not the entire U.S. population of households.
- e) No, because this would make the sample self-selected and therefore unlikely to be representative.
- f) One could stratify the sample by selecting a sample of households in each state, or by selecting urban and rural samples.
- g) One could select all of the households in each of a number of randomly chosen election districts.
- h) One could select every 10th household, by address, on each street in a city.

2
- a) A simple random sample is one that is chosen in such a way that every possible sample of the same size has an equal chance of being selected.
- b) No. This is a cluster sample. Each street has an equal chance of

being chosen. Thus there are 20 possible samples of 40 homes that each have a 1/20 chance of being selected. However, any possible sample that includes homes from several different streets has no chance of being chosen. Thus, all possible samples of size 40 do not have the same chance of being chosen.

c) Put the name of each student on a slip of paper and put the slips in a hat, then draw five names from the hat.

3

a) No, because there is no control group.

b) The placebo effect refers to the fact that some subjects may feel better just because they are taking something, not because of a specific effect of the drug.

c) Split the 50 subjects into two groups of 25, with one group receiving the drug and the other receiving a placebo.

d) Blinding for the participants in the study means that they do not know who is receiving the real drug and who is receiving the placebo.

e) An experimenter effect occurs if the experimenter somehow alerts the participants to whether they are receiving the real drug or placebo or if the experimenter decides who will get the real drug or the placebo. In the latter case, if the experimenter were to give the real drug to subjects with less serious colds, the drug might appear to be more effective than it really is. This effect can be eliminated by making the experiment double blind so that the experimenter doesn't know who is getting the drug or the placebo.

4

a) It is a convenience sample because people in Marion are more convenient than trying to reach people nationwide. It is not a cluster sample because Marion was not chosen at random; it was already known to parallel the makeup of the nation.

b) Examples: Rate the weather where you live on a scale of 1=poor to 5=great; Rate your job opportunities where you live on a scale of 1=poor to 5=great.

CHAPTER 2 ANSWERS

Section 2.1

1 This statement is sensible. Data must be either qualitative or quantitative.

3 This statement is false. While there are 25 distinct times in the data set, the times could take on any positive values; that is, they are not restricted to say, the whole numbers. So the set consists of continuous data.

5 Colors are qualitative data because they are non-numerical.

7 Waiting times are quantitative data involving measurements.

9 Days of the week are qualitative data because they are non-numerical.

11 Scores are quantitative data, representing a measure of knowledge.

13 *Yes* or *no* responses are qualitative data because they are non-numerical.

15 Federal income taxes paid by individuals are quantitative data.

17 Names of TV shows are qualitative data.

19 Numbers of defective computer components are discrete data since only integer values are possible.

21 Numbers of taxicabs are restricted to integers and are therefore discrete data.

23 Numbers of exits are discrete data since they are limited to integer values.

25 Numbers of miles traveled are continuous data since they can include fractional miles.

27 The average speeds of cars are continuous data since fractional values are possible.

29 The numbers of stars are discrete data since they are limited to integer values.

31 Average theater prices are continuous data because an average or mean can take on any value.

33 Names of parties are at the nominal level of measurement.

35 Celsius temperatures are at the interval level of measurement. These are data for which zero is an arbitrary point representing the freezing point of water.

37 Car types are at the nominal level of measurement. Car types can not be ranked or measured.

39 Safety ratings are at the ordinal level of measurement since they are qualitative data that can be arranged in a meaningful order.

41 Final course grades are at the ordinal level of measurement since they are qualitative data that can be arranged in a meaningful order.

43 City temperatures, reported in Fahrenheit or Celsius, are at the interval level of measurement. Zero is an arbitrary point on the scale.

45 Breeds of horses are at the nominal level of measurement. They can not be ranked or measured.

47 This is not meaningful since times on a clock are on an interval scale.

49 This is meaningful since speed data are on a ratio scale.

51 This is meaningful since numbers of votes are on a ratio scale.

53 This is not meaningful since SAT score data are on an ordinal scale. There is no true zero for SAT scores.

55 Numbers of cars are quantitative data, measured on a ratio scale, and are discrete. (There is a true zero and only integer values are possible.)

57 Numbers of students from each state are quantitative data, are on a ratio level of measurement, and are discrete. (There is a true zero and only integer values are possible.)

59 Times of the day limited to being on the hour are quantitative data, are on an interval level of measurement, and are discrete.

61 Finishing positions of runners are qualitative data and on an ordinal level of measurement.

Section 2.2

1 This statement is not sensible. The precision of the estimate is too high and there may be species of butterflies which have not been identified. New ones are found every year.

3 This statement is not sensible. It would be true if "relative error" were replaced by "absolute error." The astronomer could have a 1% relative error measuring something in light years, while the biologist could have a 3% relative error measuring something microscopic.

5 This statement is not sensible. No one knows the exact population of the earth at any specific time, so it is not possible to determine who is the 6 billionth person on Earth.

7 Mistakes should lead to random errors. Dishonesty should lead to systematic errors that systematically benefit the taxpayer.

9 This is a systematic error resulting in altimeter readings that are too low throughout the entire flight.

11 This is a systematic error that makes all of the weights too high by 1.2 pounds. You can correct this error by subtracting 1.2 pounds from each weight recorded during the day.

13 Random errors could be the result of mistakes in counting; systematic errors could result from classifying some pickup trucks as SUVs.

15 Random errors can result from random mistakes in the data on the returns; systematic errors can result from people intentionally omitting some income or including deductions to which they were not entitled.

17 Random errors could result from random mistakes in reading the scale; systematic errors could result from including the weight of a paper plate.

19 Random errors could result from miscounting the popped kernels; systematic errors could result from mis-classifying unpopped kernels as popped or from counting pieces of popped kernels as whole kernels.

21 The absolute error is measured value - true value = ($19.00 - $18.50) or $.50. The relative error is (measured value - true value)/true value =(19.00 - 18.50)/18.50 = 0.027 or 2.7%.

23 The absolute error is measured value - true value = $48-$65 or -$17. The relative error is (measured value - true value)/true value = (48 - 65)/65 = -17/65 = -0.26 or -26%.

25
a) These errors are random. If they were systematic, there would be a tendency for the measurements to all be too high or all be too low.
b) It is better to report the average of the 25 measurements. It is likely to be in error by less than most of the individual measurements and is more reliable than any singe measurement.
c) Systematic errors might result from a problem with the measuring device or with the definition of the "length of the room."
d) No. If there is a systematic error in the measurements, that same error will be present in the average.

27 The laser device is more precise (0.05 in. < 1/8 in.), but the tape measure is more accurate since 62.375 is closer to 62.50 than 62.90 is to 62.50. (This assumes that your actual height is what you thought it was).

29 The digital scale is more precise (0.01 kg < .5 kg), and the digital scale is more accurate since 52.88 is closer to 52.55 than 53 is to 52.55. (This assumes that your actual weight is what you thought it was).

31 No one could possibly know the exact population of the United States today, let alone in 1860, so the claim as to the exact value of the population is not believable.

33 This is a projection for the future that sounds reasonable, so the claim is believable. It is likely that the claim is based on current and projected birth and death rates (which could change). Of course, only time will tell if the projection is correct.

35 The average maximum temperature is presumably based on daily temperature measurements, which should be reliable, so the claim is believable.

37 Since no one keeps lists of all cell phone subscribers, no one could know this number so precisely; the claim is not believable.

Section 2.3

1 This statement is not sensible. The 100,000 households is not a percentage.

3 This statement is not sensible. The two comparisons use a different base. For example, if Brenda were 50 inches tall, then Ann would be 55 inches tall. To find the percentage that Brenda is shorter than Ann, we find (50 - 55)/55 = 0.0909 or 9.09%.

5 This statement is sensible. Pete's prices equal Paul's prices plus another 10%, so Pete's prices are 110% of Paul's prices.

7
a) 1/4, 0.25, 25%
b) 9/20, 0.45, 45% [.45 = 45/100 = (5)(9)/(5)(20) = 9/45]
c) 1/3, 0.33333, 33.333%
d) 23/100, 0.23, 23%

9
a) There are 450 people at the convention of whom 250 are Republicans (130 women and 120 men). Thus 250/450 = 5/9 = 0.5556 = 55.56% of the convention attendees are Republicans.
b) There are 250 women at the convention. Thus 250/450 = 55.56% of the attendees are women.
c) There are 200 Democrats at the convention. Thus 200/450 = 4/9 = 0.4444 = 44.44% of the attendees are Democrats.
d) There are 130 female Republicans at the convention. Thus 130/450 = 0.2889 = 28.89% of the attendees are female Republicans.
e) There are 80 male Democrats at the convention. Thus 80/450 = 0.1778 = 17.78% of the attendees are male Democrats.

11 The relative change is (new value - reference value)/reference value = (1483 - 2226)/2226 = -743/2226 = -0.333 or -33.3%. Thus the number of newspapers declined by 33.3% from 1900 to 1999.

13 The relative change is (new value - reference value)/reference value = (12.0 - 8.0)/8.0 = 4.0/8.0 = 0.5 = 50.0%. Thus the number of cars produced increased by 50.0% from 1950 to 1998.

15 The times must first be changed to minutes, 216 minutes in 1961 and 207 minutes in 1998. The relative change is (new value - reference value)/reference value = (179 - 216)/216 = -37/216 = -0.1713 = -17.13%. Thus the times decreased by 17.13% from 1961 to 2000.

17 The relative difference is (compared value - reference value)/reference value = (1.75 - 1.09)/1.09 = .66/1.09 = 0.606 = 60.6%. Thus the Wall Street Journal circulation is 60.6% larger than that of the New York Times.

19 The relative difference was (compared value - reference value)/reference value = (78 - 62)/62 = 16/62 = 0.258 = 25.8%. Thus O'Hare handled 25.8% more passengers in 1999 than did Heathrow. This difference could also be expressed as (62 - 78)/78 = -16/78 = -0.205 or Heathrow handled 20.5% fewer passengers than did O'Hare in 1999.

21 The relative difference can be expressed as (compared value - reference value)/reference value = (8.4 - 6.2)/6.2 = 2.2/6.2 = 0.355 = 35.5%. Thus Saudi Arabia produced 35.5% more oil than did the U.S. in 1998.

23 The relative difference can be expressed as (compared value - reference value)/reference value = (99000 - 167000) /167000 = -68000/167000 = -0.407 = -40.7%. Thus males had 47% fewer knee replacements than did females in 1998. (Women had 68.7% more than men did.)

25 If the area of Norway is 24% more than the area of Colorado, then Norway's area is <u>124%</u> of Colorado's area. (Norway's area is 100% of Colorado's area plus another 24%.)

27 If Henry earns 45% less than Ingrid, then Henry's salary is <u>55%</u> of Ingrid's salary.

29 The range of 28% to 40% can also be expressed as 34% $\pm$ 6%, so the margin of error is 6 percentage points.

31 The percentage of the world's population living in developed countries decreased by 7.6 percentage points between 1970 and 1998. This represents a relative change of (new value - reference value)/ reference value = (19.5 - 27.1)/27.1 = -7.6/27.1 = -0.280 = -28.0% or a decrease of 28.0%.

33 The five-year survival rate for blacks for all forms of cancer increased 21 percentage points between the 1960s and the 1990s. This is a relative change of (new value - reference value)/ reference value = (48 - 27)/27 = 21/27 = 0.778 = 77.8% or an increase of 77.8%.

<u>Section 2.4</u>

1 This statement is not sensible. An index is a ratio and has no units, so it is not given in dollars or any other quantity.

3 This statement is sensible. The health care quality index is used to compare the quality of health care in different states.

5 This statement is sensible. The CPI is designed to measure the overall price of consumer goods and services, so when those costs double, the CPI should approximately double as well.

7 $\text{Index} = \frac{\text{Price Today}}{\text{Price 1975}} = \frac{\$1.45}{\$.567} = 2.557 = 255.7\%$; Thus the price index number for gasoline today is 255.7 with the 1975 price as the reference value.

9 To determine the gasoline price index using the 1985 price as the reference value, divide each of the other prices by the 1985 price and express the result as a percentage. For example, the 1955 price index is found from 29.1/119.6 = 0.243 = 24.3%, resulting in a price index of 24.3. The other price indices are found similarly, dividing the price by the 1985 price.

Year	Price	Price as a Percentage of 1985 Price	Price Index (1985 = 100)
1955	29.1¢	24.3%	24.3
1965	31.2¢	26.1%	26.1
1975	56.7¢	47.4%	47.4
1985	119.6 ¢	100.0%	100.0
1995	120.5 ¢	100.8%	100.8
2000	155.0 ¢	129.6%	129.6

11 Using Table 2.1 which uses 1975 as the reference year, we see that the 1985 gasoline price index is 210.9 while the 1965 gasoline price index is 55.0. This means that the 1965 cost will be 55.0/210.9 = 0.261 times the 1985 cost or 26.1% of the 1985 cost. Thus the 1965 cost will be 0.261 x $15 = $3.92.

13 Private college costs increased by $\frac{23651-5900}{5900} = 3.01 = 301\%$, while the CPI increased by $\frac{177.1-82.4}{82.4} = 1.15 = 115\%$ during the same period.

15 The median price increased by $\frac{148000-99000}{99000} = 0.495 = 49.5\%$, while the CPI increased by $\frac{177.1-130.7}{130.7} = 0.355 = 35.5\%$ during the same period.

17 Price in Palo Alto = $\text{Price in Denver} \times \frac{\text{Index in Palo Alto}}{\text{Index in Denver}} = \$250{,}000 \times \frac{365}{87} = \$1{,}048{,}900$

Price in Sioux City = $\text{Price in Denver} \times \frac{\text{Index in Sioux City}}{\text{Index in Denver}} = \$250{,}000 \times \frac{47}{87} = \$135{,}000$

Price in Boston = $\text{Price in Denver} \times \frac{\text{Index in Boston}}{\text{Index in Denver}} = \$250{,}000 \times \frac{182}{87} = \$523{,}000$

19 Price in Spokane = $\text{Price in Cheyenne} \times \frac{\text{Index in Spokane}}{\text{Index in Cheyenne}} = \$250{,}000 \times \frac{78}{75} = \$260{,}000$

Price in Denver = $\text{Price in Cheyenne} \times \frac{\text{Index in Denver}}{\text{Index in Cheyenne}} = \$250{,}000 \times \frac{87}{75} = \$290{,}000$

Price in Juneau = $\text{Price in Cheyenne} \times \frac{\text{Index in Juneau}}{\text{Index in Cheyenne}} = \$250{,}000 \times \frac{100}{75} = \$333{,}300$

Chapter Review Exercises

1
a) The number of people who died was 2223 x .3176 = 706
b) The number of survivors is discrete since it must be an integer.
c) The fraction of Titanic passengers who were women or children was 531/2223 = 0.2389. Thus 23.89% of the passengers were women or children.
d) The number of boys is 45 + (.42 x 45) = 45 + 19 = 64.
e) Ages have a ratio level of measurement since there is a true zero.
f) The categories of passengers are at the nominal level of measurement.

2
a) The salary is very precise since it is given to the nearest penny.
b) It is based on a small sample and can only approximately represent the average salary of the population. Thus it should not be stated so precisely.
c) Next year's projected income level is $43,782.64 x (1 + 0.04) = $45534.
d) The suspected error is systematic because the lack of low-income respondents systematically makes the average too high.
e) The absolute error is (measured value - reference value) = 39376.69 - 43782.64 = -4405.95. Thus the initial value is too high by $4405.95. If the initial larger value is used as the reference value, the relative error is (measured value - reference value)/reference value = -4405.95/43782.64 = -0.103 or -10.3%. If the later correct value is used as the reference value, the relative error is 4405.95/39376.69 = 0.112 or 11.2%.

3
a) $\text{Index for 1996} = \frac{\text{Price in 1996}}{\text{Price in 1994}} = \frac{396}{375} = 1.056 = 105.6\%$. Thus the index for 1996 is 105.6.
b) $\text{Index for 2001} = \frac{\text{Price in 2001}}{\text{Price in 1994}} = \frac{297}{375} = 0.792 = 79.2\%$. Thus the index for 1998 is 93.9.
c) The absolute change is 297 - 317 = -20. The relative change is -20/317 = -0.06 = -6%.
d) The years are on an interval level of measurement since there is no true zero.
e) The amounts of barley are on a ratio level of measurement since there is a true zero.

CHAPTER 3 ANSWERS

Section 3.1

1 This statement is not sensible since a frequency table must have a column of counts (numbers of occurrences).

3 This statement is not sensible. Relative frequencies must be between zero and one.

5 This statement is sensible since the data is spread over the same range regardless of the number of bins. If the bins are made narrower, there have to be more of them to cover the entire range of the data.

7

Grade	Frequency	Relative Frequency	Cumulative Frequency
A	4	0.167	4
B	7	0.292	11
C	8	0.333	19
D	3	0.125	22
F	2	0.083	24
Total	**24**	**1.000**	**24**

The Frequency column lists the number of times each grade occurred. The Relative Frequency column shows the portion of the time each grade occurred and the entries are obtained by dividing the Frequency by the total of 24. The Cumulative Frequency column gives the total number of grades that are equal to or better than the grade shown in the row.

9 a)

Jump Length (Feet)	Frequency	Relative Frequency	Cumulative Frequency
21-21.99	6	0.500	6
22-22.99	5	0.417	11
23-23.99	1	0.083	12
Total	**12**	**1.000**	**12**

b)

Jump Length	Frequency	Relative	Cumulative
20-21.99	6	0.50	6
22-23.99	6	0.50	12
Total	**12**	**1.00**	**12**

11

Weight (Pounds)	**Frequency**	**Relative Frequency**	**Cumulative Frequency**
0.7900-0.7949	1	1/36	1
0.7950-0.7999	0	0	1
0.8000-0.8049	1	1/36	2
0.8050-0.8099	3	3/36	5
0.8100-0.8149	4	4/36	9
0.8150-0.8199	17	17/36	26
0.8200-0.8249	6	6/36	32
0.8250-0.8299	4	4/36	36
Total	**36**	**1**	**36**

13 For Categories B and C, determine the frequencies by multiplying the total, 50, by 18% and 24% respectively, yielding frequencies of 9 and 12. Then add the frequencies for categories B, C, D, and F; subtract this total of 38 from the overall total of 50 to get the frequency for category A, which is 12. Now obtain each of the remaining relative frequencies by dividing the frequency by the total of 50 and converting to a percentage. For example, 12/50 = 0.24 = 24%.

Category	**Frequency**	**Relative Frequency**
A	12	0.24
B	9	0.18
C	12	0.24
D	11	0.22
F	6	0.12
Total	**50**	**1.00**

15 As you can see from the table, about 1/3 of the stocks outperformed a bank account with 3% annual return (10 out of 29).

Total Return in 2000	Frequency (Number of Companies)
-70 to -60.1%	1
-60 to -51.1%	0
-50 to -40.1%	0
-40 to -30.1%	3
-30 to -20.1%	7
-20 to -10.1%	4
-10 to -0.1%	3
0 to 9.9%	3
10 to 19.9	2
20 to 29.9	3
30 to 39.9	0
40 to 49.9	1
50 to 59.9	0
60 to 69.9	1
70 to 79.9	0
80 to 89.9	0
90-99.9%	0
100% or more	1
Total	29

17 As seen from the table, most awards for actors are won when the actors are in their 30s or 40s with none winning while in their 20s. Actresses, however, have won as many awards in their 20s as in their 40s with the peak number of awards coming in their 30s (See Example 3 in Section 3.2).

Age	Number of Actors
20-29	0
30-39	12
40-49	13
50-59	5
60-69	3
70-79	1

19

Tax	Frequency	Relative
$500-999	1	0.02
$1000-	8	0.16
$1500-	31	0.62
$2000-	7	0.14
$2500-	3	0.06
Total	50	1.00

Most of the states have per capita state taxes that fall in the middle of the range. The highest per capita state tax is more than three times the lowest one. (Note that you could have made other choices for categories with a width of 500. For example, you could have started with $800-1299.)

21

Teacher Salary	Frequency	Relative Frequency
$24000-27999	3	0.06
$28000-31999	10	0.20
$32000-35999	15	0.30
$36000-39999	9	0.18
$40000-43999	7	0.14
$44000-47999	1	0.02
$48000-51999	5	0.10
Total	50	1.00

The results in this table somewhat parallel the results in the table of Exercise 15, with most of the states having average teacher salaries in the range from $28000 to $43999. The highest state average salaries are almost twice what the lowest ones are.

23 a) Ages are quantitative continuous data, and transportation is qualitative nominal data.

b)

	Transportation					
Age		1	2	3	4	5
	1	2	1	1	1	0
	2	1	2	0	0	0
	3	1	0	1	1	2
	4	0	0	1	1	0
	5	1	1	0	0	3

Section 3.2

1 This statement is not sensible. Because the categories (ages) are numerical and ordered, a histogram is needed.

3 This statement is not sensible. One variable in a time series diagram must be time.

5 This statement is tasty, but not sensible. A time-series diagram would be used for this purpose.

7 a) Bar graph

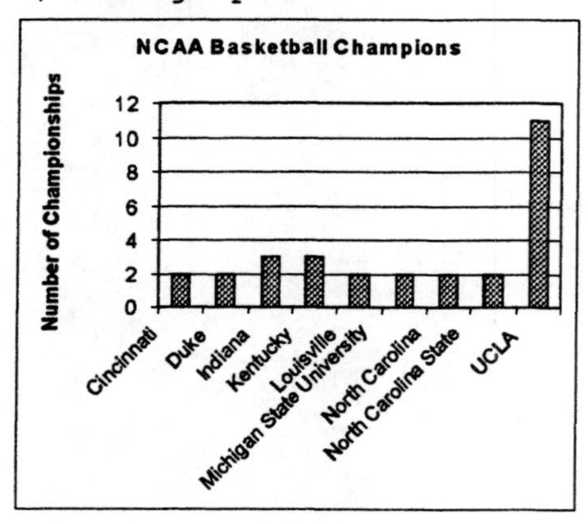

c) Pareto Chart

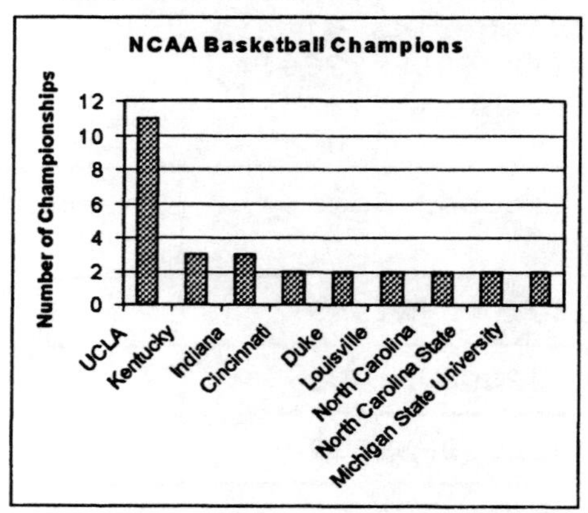

b) Dot plot

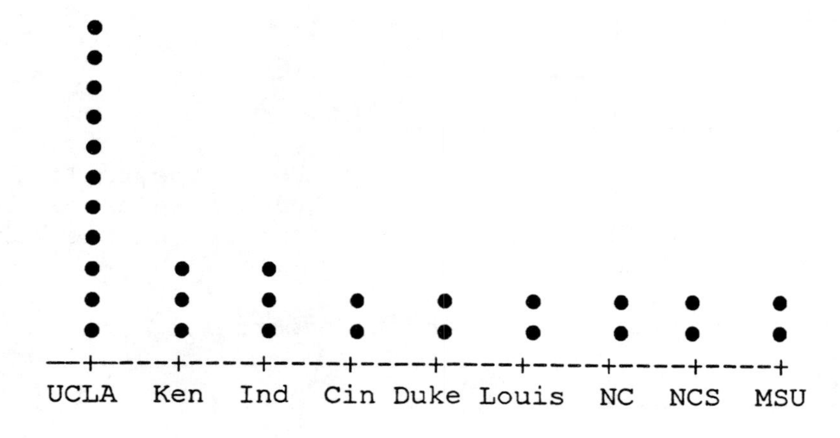

9 a) Bar graph

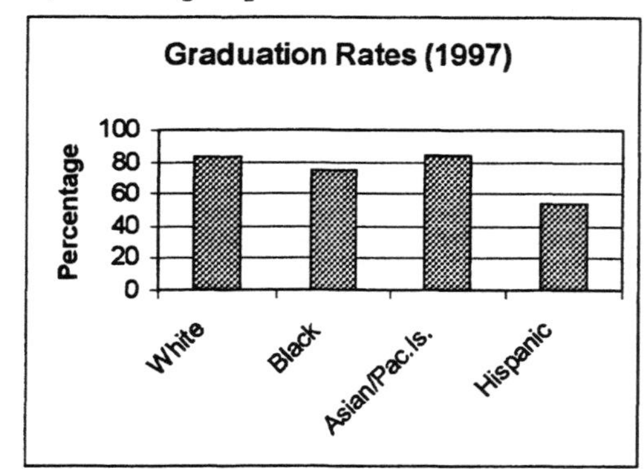

b) Pareto Chart

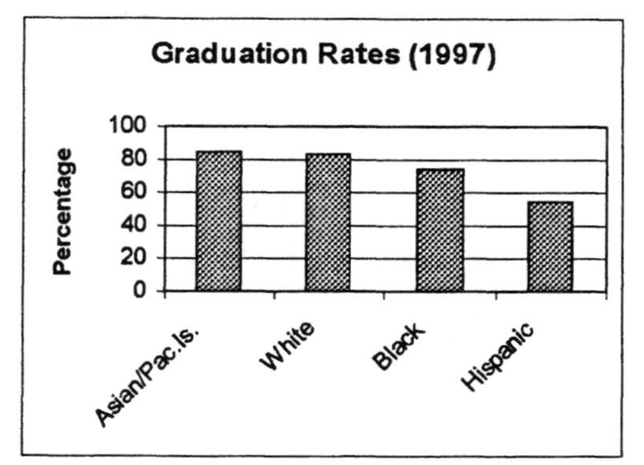

c) Since the data indicate that the lowest graduation rates are among the black and Hispanic populations, we might want to first look at designing a program that is primarily intended for these ethnic groups. Of course, other data might very well be relevant and should be analyzed. Is there a geographical explanation for the data? Are the problems primarily in states that have high populations of these groups? Are there areas where whites and Asians and Pacific Islanders also have low graduation rates? Is there an urban/rural element to the problem?

11 a)

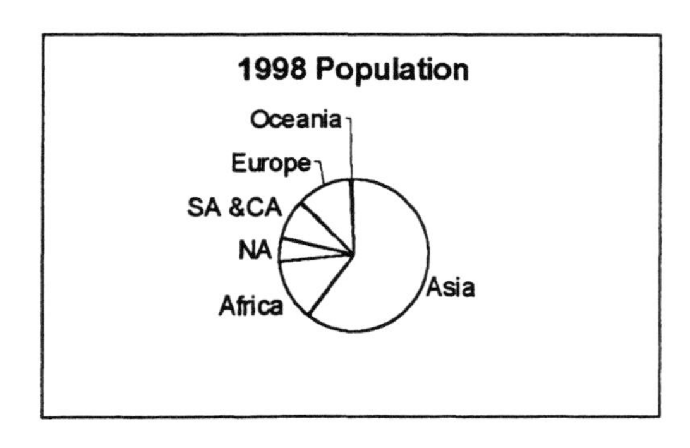

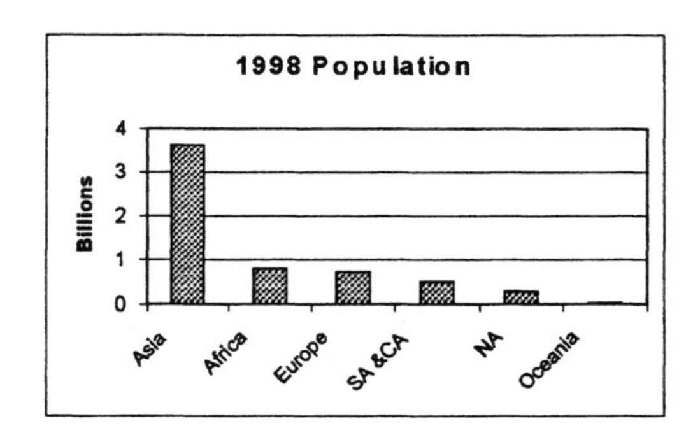

The pie chart is designed to show percentages, but does not convey the actual population. It does, however, demonstrate the enormous percentage of the world population that resides in Asia. The Pareto chart is better for conveying the actual population sizes of the various continents.

b) A line chart would not be appropriate because the categories are at the nominal level of measurement.

13

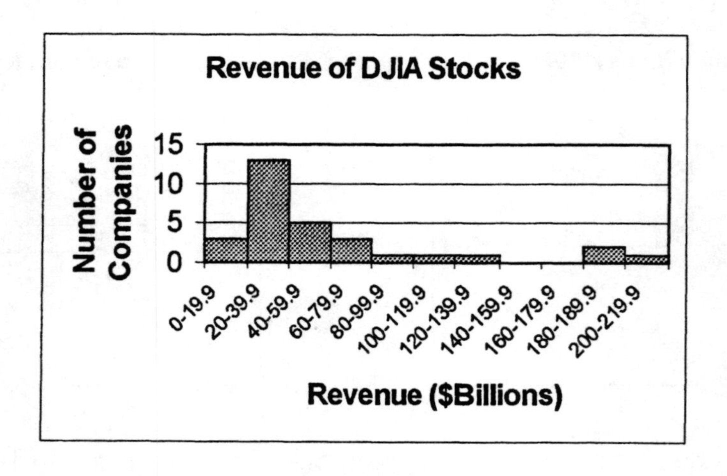

15 a)

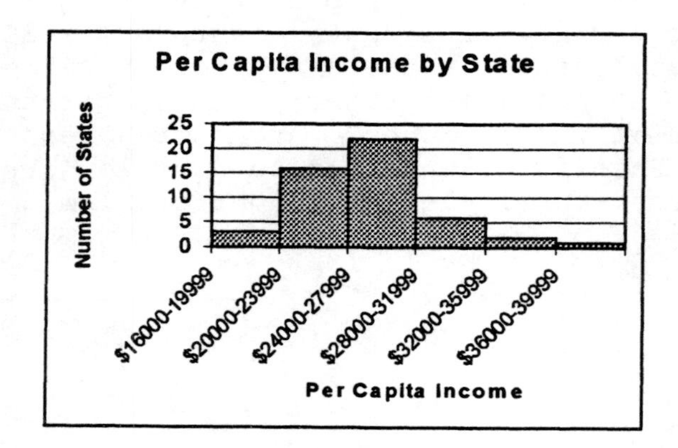

b)

```
$16000-19999 MS WV NM
$20000-23999 MT AR UT OK ID SC LA AL KY ND SD ME AZ WY TN IA
$24000-27999 NC VT IN MO NE OR TX KS GA WI OH AK FL MI HI PA RI NV VA CA MN WA
$28000-31999 CO IL NH DE MD NY
$32000-35999 MA NJ
$36000-39999 CT
```

c) There is a fairly wide range of per capita incomes for the various states. Note, in particular, that the highest incomes are about twice the size of the lowest income. We also see that seven of the nine high income states are in the Northeast.

17 a)

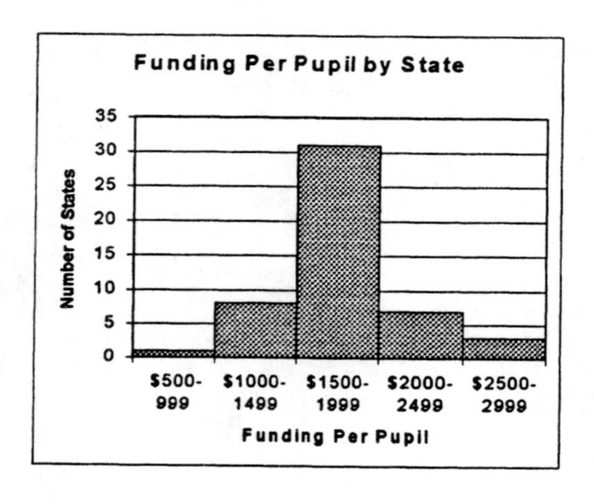

b)

$3500-3999 UT
$4000-4499 AZ AR OK MS
$4500-4999 ID LA ND SD TN NM MO NV
$5000-5499 NC SC CO NE AL CA KY MT IL FL KS
$5500-5999 OH TX GA HI IA WA IN VA OR
$6000-6499 NH WY MN ME WV
$6500-6999 VT WI MD MI PA
$7000-7499 MA DE
$7500-7999 RI
$8000-8499 CT
$8500-8999 NY AK
$9000-9499 NJ

c) There is a wide range of expenditures per pupil. While 38 states expend $3000 to $7000 per pupil, the highest spending state invests about 2.5 times as much in education as the lowest spending state.

19

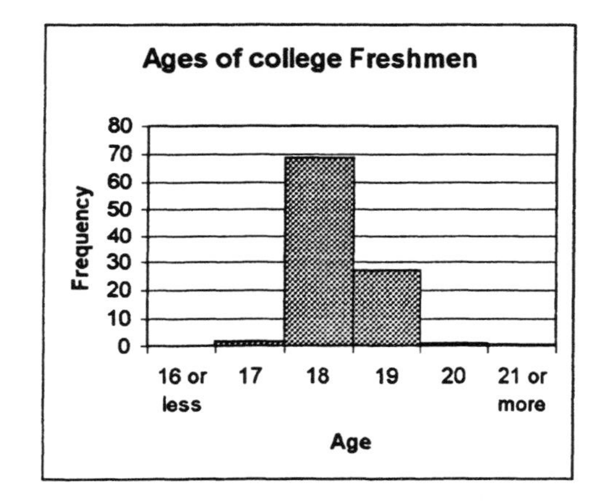

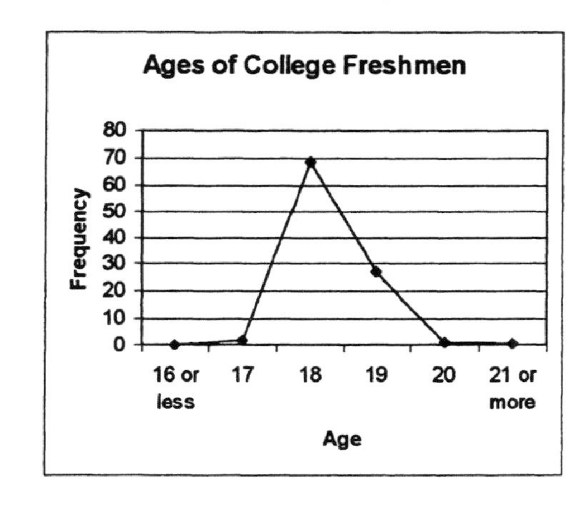

21

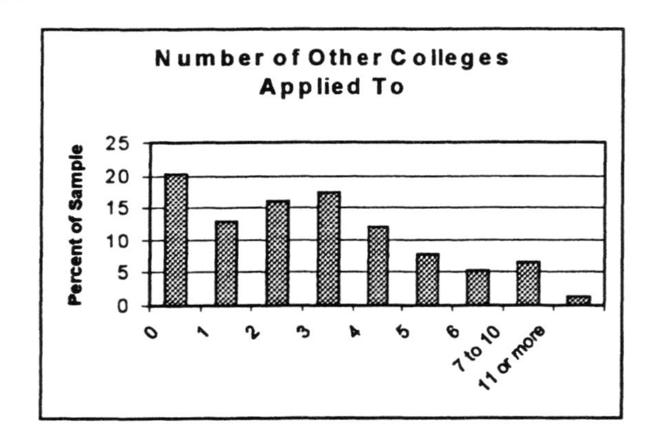

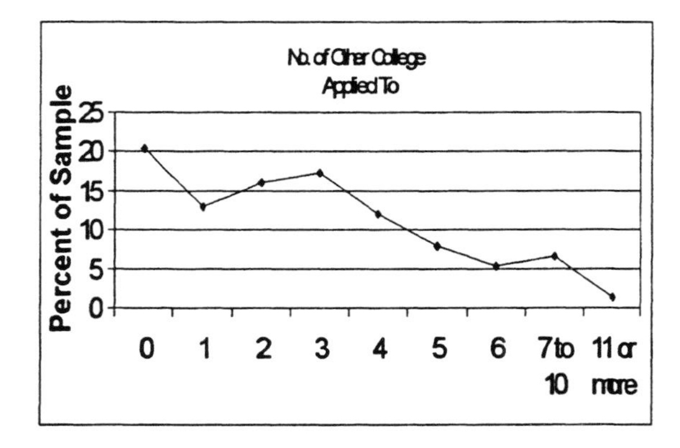

23 a) 25,000 in 1982; 16,652 in 2000. There is a decreasing overall trend in alcohol-related fatalities over this period.

b) We use the 1982 figure of 25,000 fatalities as the base for reference. Therefore the change in alcohol-related fatalities is (16652-25000)/25000 = -8348/25000 = -0.334 or a decrease of about 33%.

c) In 1982, 25000/43945 = 0.57 or 57% of all fatalities involved alcohol. In 2000, 16652/41821 = 0.398 or 40% of all fatalities involved alcohol.

d) A number of factors probably contribute to the decrease in the percentage of accidents that involve alcohol. These include more and better education programs about the effects of driving after drinking, non-drinking driver promotions, stiffer penalties for drunk driving, lower blood alcohol requirements in many states for being judged drunk, and, more recently, drivers notifying the police by cell phone about others operating under the influence.

25 A Pareto chart or pie chart would work well. Since the number of colors is likely to be large, a Pareto chart will probably be preferred and will make it easy to pick out the most preferred colors.

27 A time-series line chart is best so that the prices are shown by the year.

Section 3.3

1 This statement is not sensible. A two-dimensional time series diagram would suffice.

3 This statement is not sensible. A contour map is used to display a continuous variable, such as temperature or elevation. The Walmart data consists of a single number for each state.

5 a) Teen smoking rates have generally risen for all ages, with a light decline from 1996 to 1997 among 8th and 10th graders. Also, in each year, there are more 12th graders smoking than 10th graders and more 10th graders smoking than 8th graders.

b)

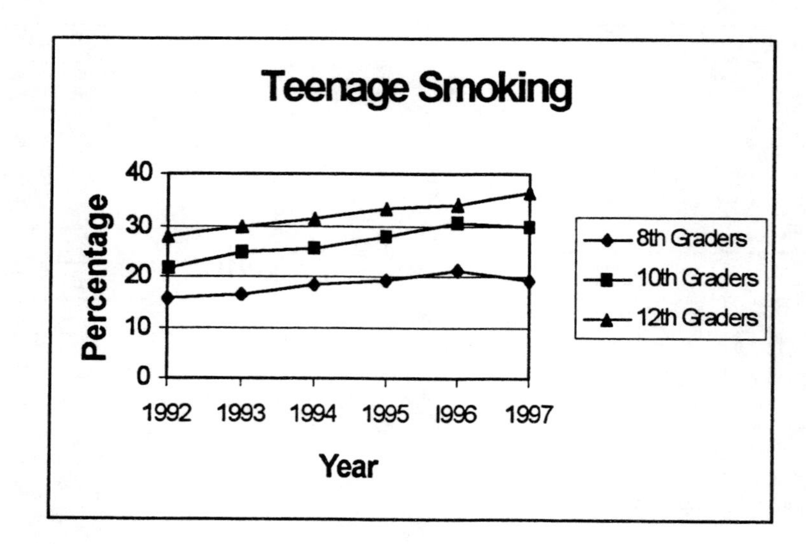

The line graph makes it easier to judge the trend over time. It also has the appearance of being less cluttered and can therefore be interpreted and understood more quickly.

7 a) The purpose of the graph is to show that earnings for people with college and advanced degrees have increased more rapidly than those with less education. The three sets of bars represent the

three years 1975, 1985, and 1995.

b) The 3-dimensional appearance is cosmetic and makes the values somewhat more difficult to read.

c) For people with Bachelor's degrees, earnings rose from about $10,000 to more than $30,000. For people who did not graduate from high school, earnings rose from about $6,000 to about $12,000.

9 a) Ireland is represented by the top box in the stack. Subtracting the lower level from the upper level, we see that about 1.7 to 1.8 million Irish immigrants arrived during 1841-1860.

b) Subtracting the lower level of the Canada box from the upper level, we see that about 0.9 million Canadian immigrants arrived during 1901-1920.

c) Subtracting the lower level of the Mexico box from the upper level, we see that about 3.9 million Mexican immigrants arrived during 1981-2000.

d) Total immigration increased steadily from 1821 to 1920, with most of the immigrants arriving from European countries. During the depression and World War II years of the 1920s, 1930s, 1940s and into the 1950s, immigration decreased substantially. For the last 40 years, immigration has increased rapidly, but there has been a shift in the countries of origin to Mexico, the Philippines, and for the last ten years or so, China. The reasons for this shift probably include the better education and job opportunities in the U.S. compared to those in the countries of origin.

11 a) 1930: men, 75,000; women, 50,000.
2000: men, 500,000; women, 650,000.

b) In 1980, slightly more men than women earned degrees; In 2000, more women than men earned degrees.

c) The total number of degrees increased the most during the 1960s. This is the period when the upper line of the graph has the steepest upward slope.

d) In 1950, about 450,000 degrees were awarded; in 2000, there were about 1,150,000.

e) The stack plot makes it easy to follow the trend for degrees awarded to men and the trend for total degrees awarded. However, it makes a direct comparison of the number of degrees awarded to men and women a little difficult. A line graph or a double bar chart showing the separate data for men and women would be more effective for comparison purposes.

13

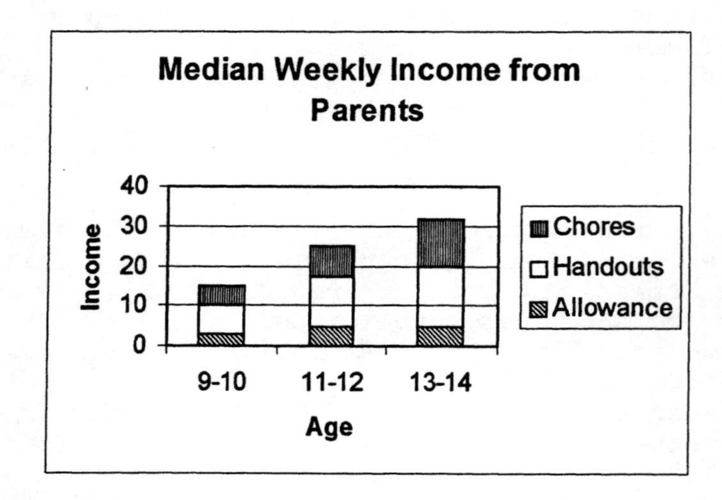

The side-by-side bars of Figure 3.17 make it easier to compare the three sources of funds for the three age groups. The stacked bars of this exercise make it easier to see the cumulative effect of all three sources of funds for each age group.

15 There are significant regional differences. For example, the probability that a black student has white classmates is generally much higher in the north than in the south and in rural areas than in urban areas.

17 A Pareto chart for the men's data and another one for the women's data will best point out the most common types of cancer that caused deaths to each gender during 2000. For most of those types of cancer common to both genders, the order of importance and the numbers of deaths are similar for men and women. Men suffer significantly more lung cancer deaths than women, but lung cancer is the most prevalent cancer causing death for both.

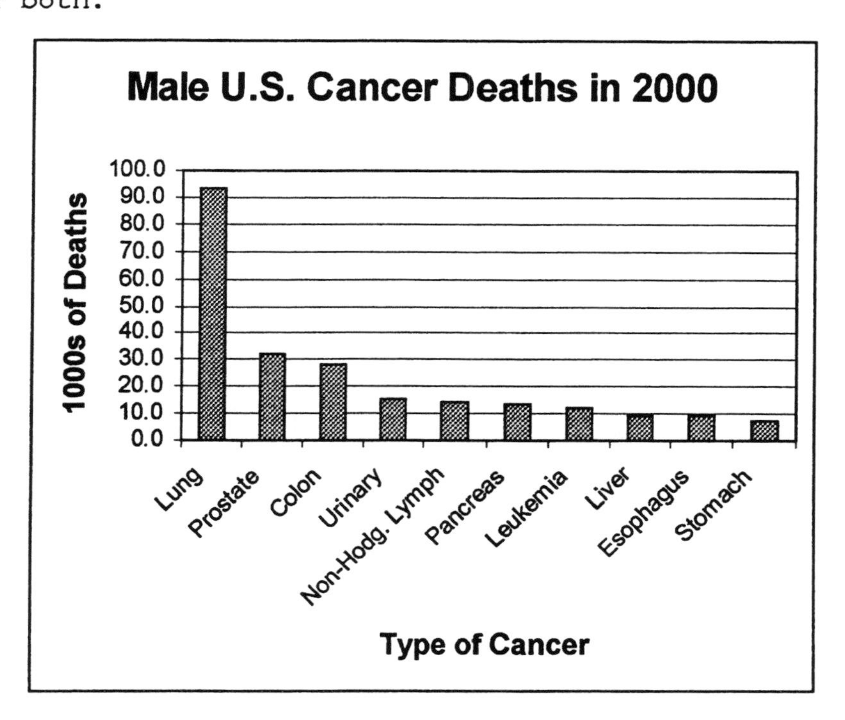

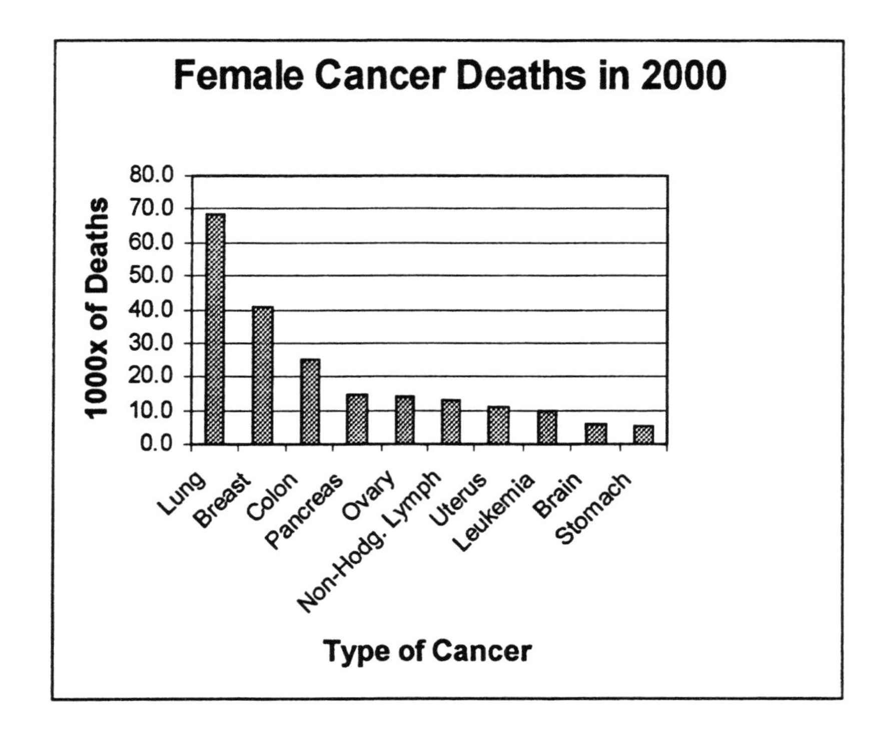

19 One way to display these data is to use 2 line graphs, one for the number of daily newspapers and one for the circulation. The horizontal axis will contain the years. The line for the number of newspapers will show a steady decline, while the circulation increased until around 1990, and then dropped off. This drop-off may be related to the increasing use of the internet and the availability of all kinds of news there. One could also use a multiple line graph, with different scales on two Y-axes for number of newspapers and circulation (not shown).

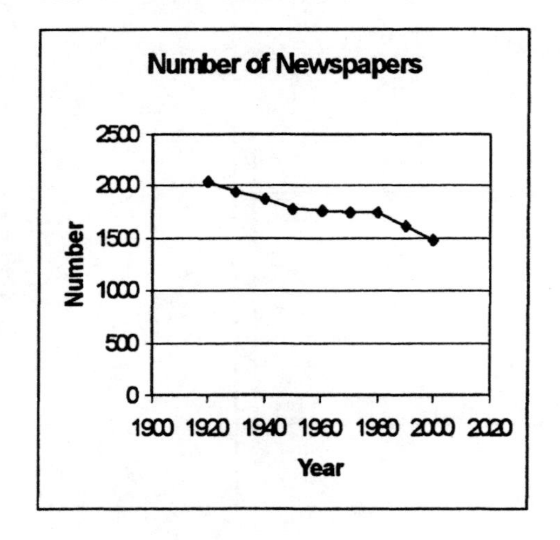

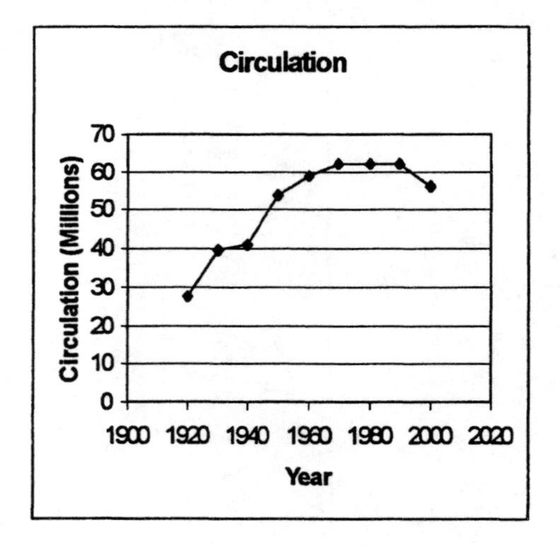

Section 3.4

1

a) Because of the 3-d appearance of the pie charts, the sizes of the wedges on the page do not match the percentages. Instead, they show how the wedges would look if the entire pie were tilted at an angle. This distortion makes it difficult to see the true relationships among the categories.

b) Because of the 3-d appearance of the pie charts, the sizes of the wedges on the page do not match the percentages. Instead, they show how the wedges would look if the entire pie were tilted at an angle. This distortion makes it difficult to see the true relationships among the categories.

c) The pie charts below are not distorted by a 3-d view.

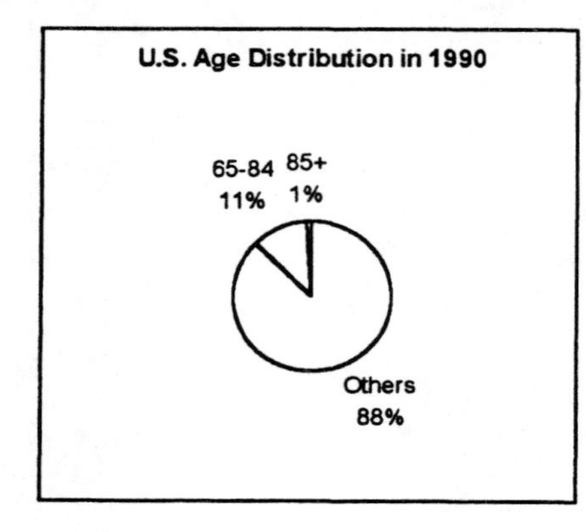

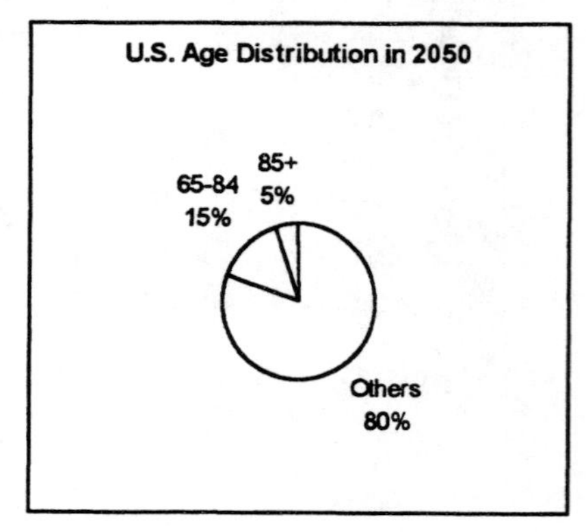

d) Both the 65-84 and 85+ populations are expected to be a larger percentage of the total population in 2050 than they were in 1990.

3 The vertical scale does not begin at zero, which in this case exaggerates the difference in pay between men and women. Here is the same graph drawn more fairly.

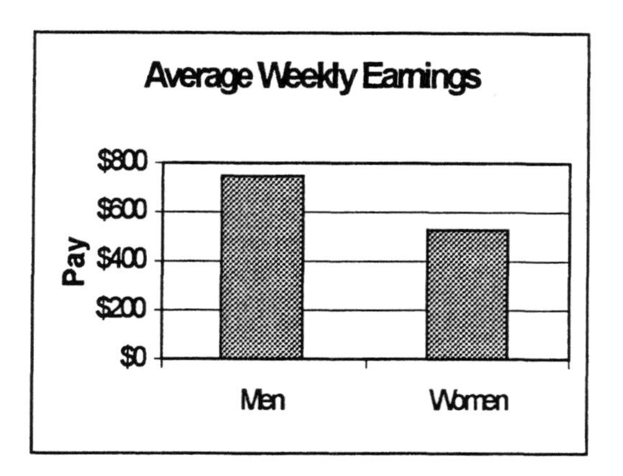

5

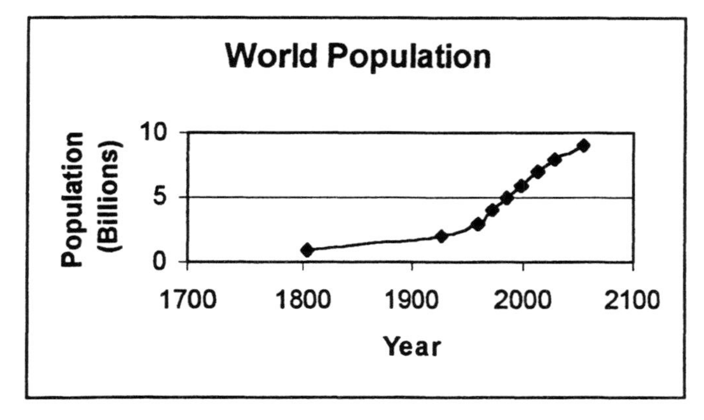

With the graph recast, we see the dramatic rise in world population that has occurred in recent years.

7 a) If we start the vertical scale at zero, the changes look much smaller.
b) A horizontal (flat) line is consistent with the error bars, so the claim of seasonal variation is not justified by this graph.

9 The percent change in the CPI was greatest in 1990. The change was least in 1986. Prices have increased in every year, but the increases in the latter 1990s have generally been smaller than those over the previous several years.

11 The first baby boom peaked in the late 1950s and the second peaked around 1990. The designers of the graph probably intended to show the similarity in the rise and fall of the number of births over the two periods which are each about one generation (18 years) in length.

Chapter Review Exercises

1 a) Sunday Rain

Rainfall Amount	Frequency
0.00-0.09	44
0.10-0.19	0
0.20-0.29	5
0.30-0.39	1
0.40-0.49	1
0.50-0.59	0
0.60-0.69	0
0.70-0.79	0
0.80-0.89	0
0.90-0.99	0
1.00-1.09	0
1.10-1.19	0
1.20-1.29	1

b) Wednesday Rain

Rainfall Amount	Frequency
0.00-0.09	45
0.10-0.19	3
0.20-0.29	1
0.30-0.39	1
0.40-0.49	0
0.50-0.59	0
0.60-0.69	2

c) Both tables have the highest frequency for low values, but the data vary over a greater range for the Sunday rainfall than for the Wednesday rainfall.

2 a)

Rainfall	Relative Frequency
0.00-0.09	44/52
0.10-0.19	0/52
0.20-0.29	5/52
0.30-0.39	1/52
0.40-0.49	1/52
0.50-0.59	0/52
0.60-0.69	0/52
0.70-0.79	0/52
0.80-0.89	0/52
0.90-0.99	0/52
1.00-1.09	0/52
1.10-1.19	0/52
1.20-1.29	1/52
Total	1

b)

Rainfall Amount	Cumulative Frequency
0.00-0.09	44
0.10-0.19	44
0.20-0.29	49
0.30-0.39	50
0.40-0.49	51
0.50-0.59	51
0.60-0.69	51
0.70-0.79	51
0.80-0.89	51
0.90-0.99	51
1.00-1.09	51
1.10-1.19	51
1.20-1.29	52

3 a)

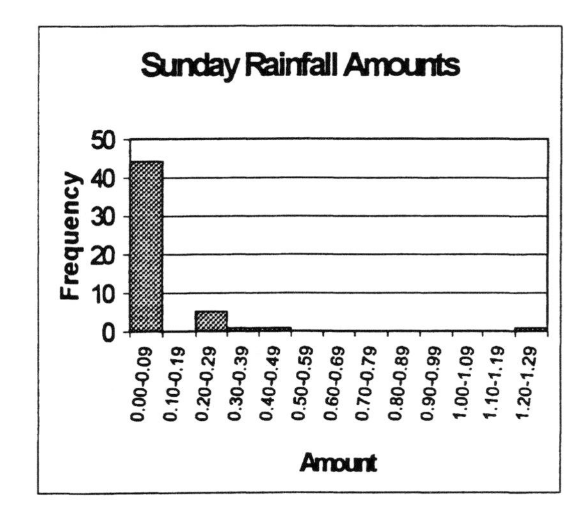

b)

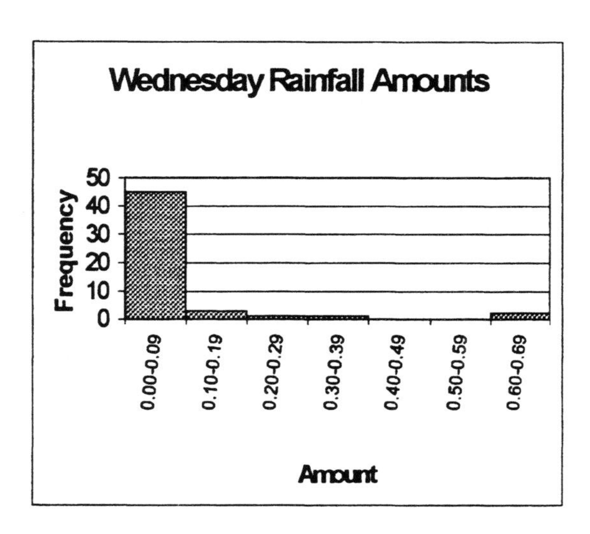

c) The histograms visually show the same trends that we saw in the tables of Exercise 1. (Note that the horizontal scales in the two histograms above are not the same, but it is clear from the labeling on the axis that there is more variation in the Sunday data.)

4

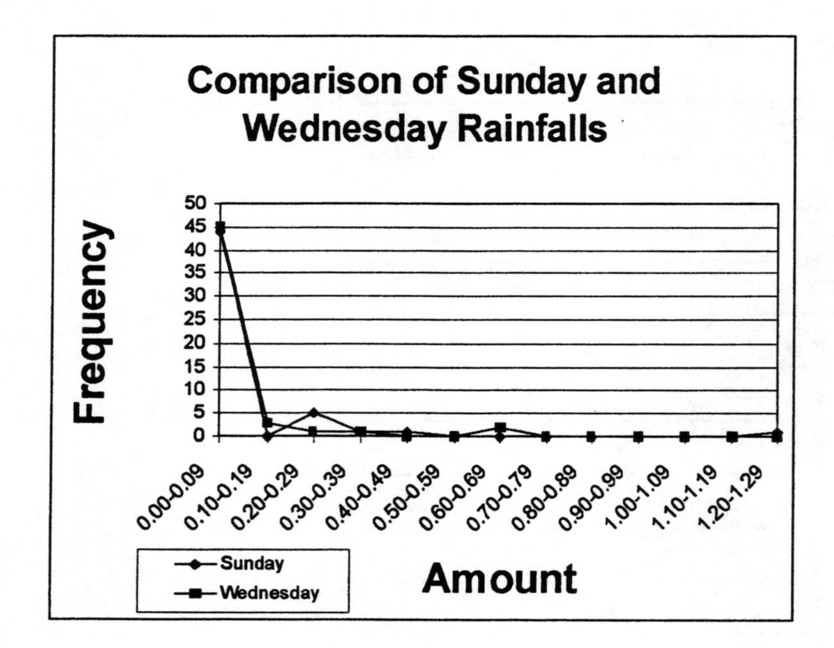

Based on the histograms in Exercise 3 or on the line chart above, there does not appear to be an appreciable difference between weekend and weekday rainfall, with the exception of the one high rainfall day that occurred on a Sunday.

5

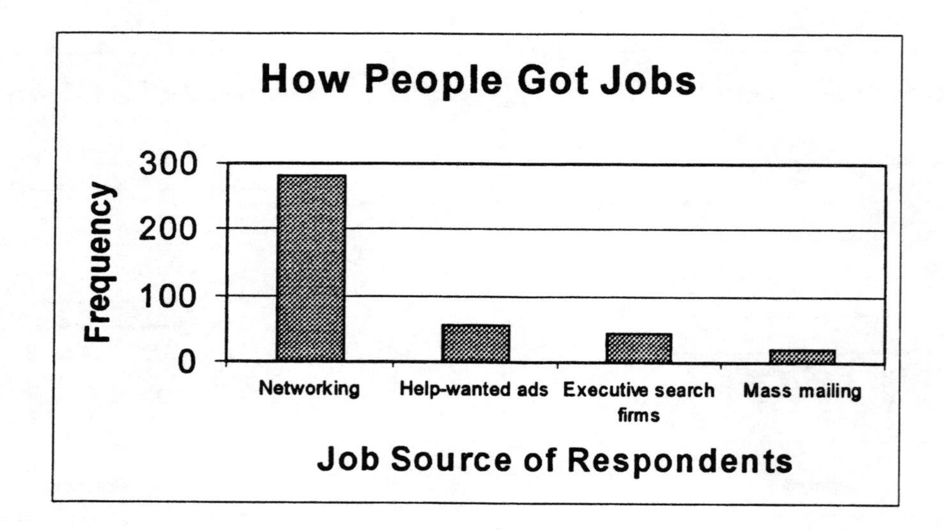

Networking appears to be the best approach to finding a job. At least, it was the method used most often by successful job seekers. One would need additional data from job seekers who were not successful to determine the success rate for each of the four methods.

6

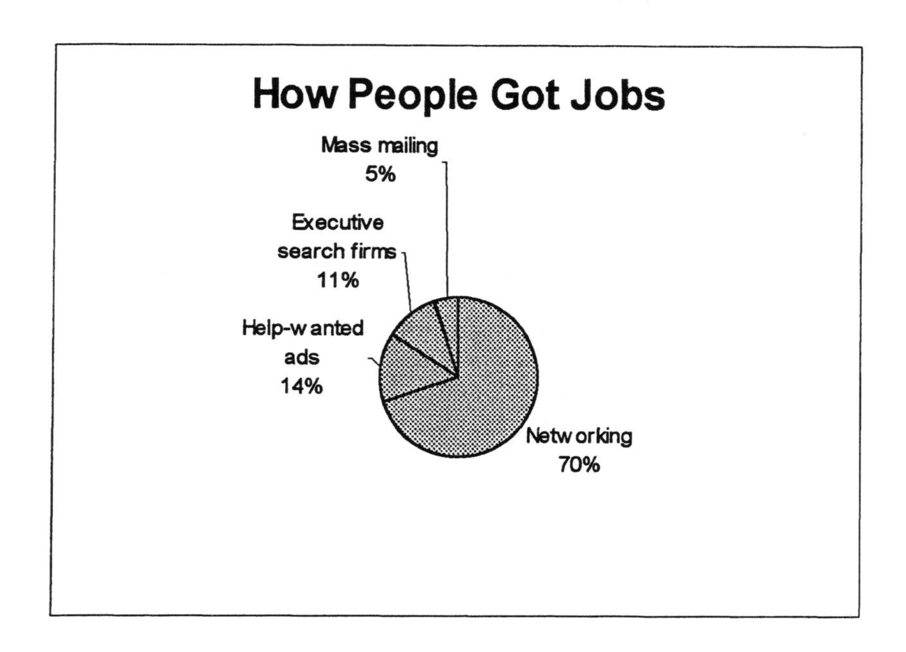

The pie chart is more useful for determining the *relative* importance of the job sources for successful job seekers.

CHAPTER 4 ANSWERS

Section 4.1

1 This statement is not sensible. The mean may be greater than or less than the mode in any given data set.

3 This statement is sensible. An outlier on the heavy side would raise the mean above the median.

5 This statement is not sensible. A data set has one mean.

7 $$\text{Mean} = \frac{672.2+679.2+669.8+672.6+672.2+662.2}{6} = \frac{4028.2}{6} = 671.37$$

For the median, first put the six numbers in increasing order. Since there is an even number of data values, the median is the average of the middle two (third and fourth) numbers. Thus median = (672.2 + 672.2)/2 = 672.2.
The mode is the number that occurs most often. Since 672.2 occurs twice and no other number occurs more than once, 672.2 is the mode.

9 $$\text{Mean} = \frac{.27+.17+.17+.16+.13+.24+.29+.24+.14+.16+.12+.16}{12} = \frac{2.25}{12} = 0.188$$

For the median, first put the twelve numbers in increasing order. Since there is an even number of data values, the median is the average of the middle two (sixth and seventh) numbers. Thus median = (0.16 + 0.17)/2 = 0.165.
The mode is the number that occurs most often. Since 0.16 occurs three times and no other number occurs more than twice, 0.16 is the mode.

11 $$\text{Mean} = \frac{15+11+10+9+0+2+4+5+5+7+10+12+15+18+19}{15} = \frac{142}{15} = 9.5$$

For the median, put the fifteen numbers in increasing order. Since there is an odd number of data values, the median is the middle (eighth) number. Thus the median = 10.
The mode is the number that occurs most often. Since 5, 10, and 15 each occur twice and no other number occurs more than once, there are three modes (5, 10, and 15).

13 $$\text{Mean} = \frac{.957+.912+.842+.925+.939+.886+.914+.913+.958+.947+.920}{11} = \frac{10.113}{11} = 0.9194$$

For the median, put the eleven numbers in increasing order. Since there is an odd number of data values, the median is the middle (sixth) number. Thus the median = 0.920.
The mode is the number that occurs most often. Since all numbers occur once, there is no mode.

15 a) Mean = total area/number of states = 1103100/7 = 157,586 acres
Median = 104,100 (the fourth smallest in the ordered list of seven)
b) Alaska is the outlier on the high end. If Alaska is removed from the list, the new mean is 487900/6 = 81,317 acres. The median becomes (53,200 + 104,100)/2 = 157,300/2 = 78,650 acres.
c) Connecticut is the outlier on the low end. If Connecticut is

removed from the list, the new mean is 1,097,600/6 = 182,933 acres. The median becomes (104,100 + 114,000)/2 = 218,100/2 = 109,050 acres.

17 a) $$\text{Mean} = \frac{70+75+80+70}{4} = \frac{295}{4} = 73.75$$

b) Since the mean will be the total divided by five, the total will need to be 5 x 75 =375 in order for the mean to be 75 after the next quiz. Since the total is already 295 after the first four quizzes, the next quiz will need to be 375 - 295 = 80.

c) If you achieve a score of 100, your mean score will be (295+100)/5 = 395/5 = 79. Thus it's not possible to have a mean score higher than 79 after the next quiz.

19 Since the mean equals the total divided by 6, you must have a total of 480 in order to have a mean of 80. If you get 90 on the next quiz, you will have a total of 570 for seven quizzes for a mean of 570/7 = 81.4. The maximum mean score that you could have after the next quiz would result if you scored a 100. This would make your total 580 and your mean would be 580/7 = 82.9. The minimum mean score that you could have after the next quiz would result if you scored a zero. In that case, your new mean would be 480/7 = 68.6.

21 The number of hits that she has so far is 30 x .300 = 9. If she gets a hit in her next at-bat, she will have 10 hits in 31 at-bats. Her new batting average will be 10/31 = .323.

23 The mean height (in inches) of your players is (77+78+78+84+86)/5 = 403/5 = 80.6" or 6' 8.6". The median height is 78" or 6' 6". The answer to the question depends on the meaning of "average." If the "average" height reported by the league is a mean, your team is above average height; if it is a median, the team is below average height.

Note: The solutions provided below for Exercises 25, 27, and 29 are very subjective, and a different solution might be more appropriate if the purpose of knowing the "average" were known to be something specific.

25 Since oranges packed in a large box are usually pre-sorted so that they are similar in size, either the mean or the median will provide a good average. If the box contains a random collection of oranges just picked, the mean would be a better (and quicker) average to use since only the total weight and total number of oranges are needed to find it.

27 The mean would best describe the average number of pieces of lost luggage per flight. There will likely be very few large numbers to influence the value of the mean and the mean also reflects the total number of pieces lost.

29 This average depends both on how many items people have bought and on the number of checkout lines in operation. The median would not be influenced by a few people with "mountains" of groceries and would provide a "typical" waiting time. The mean would reflect the total waiting times of all customers and might provide valuable information for a store manager.

31 No. The classes are not of equal size. If we think of the two percentages as points out of 100, then the first class had a total number of points equal to 25 x 86 = 2150 while the second had 30 x 84 = 2520. The mean for the two classes combined is the total number of points divided by the total number of students or 4670/55 = 84.91.

33 This requires a weighted mean of the grades where the weights are the percentages. Therefore,

$$\text{Mean} = \frac{(15)(75)+(20)(90)+(40)(85)+(25)(72)}{15+20+40+25} = \frac{8125}{100} = 81.25$$

35 No. Suppose that the player had 400 hits in 1000 at-bats (.400 average) followed by 2 hits in 4 at-bats. The player now has 402 hits in 1004 at-bats for an average of .4003 (which would still be reported as a .400 average).

37 a) $$\text{Batting Average} = \frac{\text{Total number of hits}}{\text{Total At-bats}} = \frac{3+2+2}{5+4+5} = \frac{7}{14} = .500$$

b) $$\text{Slugging Average} = \frac{\text{Total number of bases}}{\text{Total At-bats}} = \frac{3+4+6}{5+4+5} = \frac{13}{14} = .929$$

c) Yes. For example, if a player has 2 home runs in 4 at-bats, the slugging percentage is 8/4 = 2 or 200%.

39 This is a weighted mean with the course credits being the weights.

Thus, $$\text{Mean} = \frac{(5)(4)+(3)(3)+(3)(2)+(3)(1)}{5+3+3+3} = \frac{38}{14} = 2.71$$

Section 4.2

1 This statement is not sensible. A distribution can have any number of modes and be symmetric.

3 This statement is sensible. Weights of dogs of the same breed should have less variation than weights of dogs of several breeds.

5 This statement is not sensible. With a symmetric distribution, the mean, median, and mode are equal.

7 This distribution has one mode (at 1 month), is right skewed, and has moderate variation.

9 a) The distribution of incomes will have a shape similar to the one shown below, but its exact shape cannot be determined from the information given. Since the mean is greater than the median, it will be right-skewed.

b) About 50% or 150 (half of 300) of the families earned less than $35,000 since that is the value of the median.

c) No. It depends on the precise distribution.

Frequency

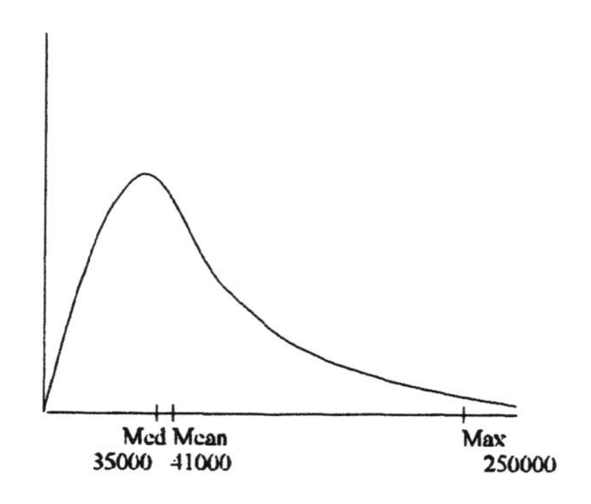

Income

11 a) The distribution is likely to have one mode.
b) The distribution is likely to be nearly symmetric, perhaps a little left-skewed since there is a possible range of 75 points below the mean and only 25 points above it.
c) There will be moderate variation.

13 a) The distribution is likely to have one or two modes; there might be one mode for linemen, fullbacks, and linebackers, and a second mode for running backs, wide receivers, and defensive backs.
b) The distribution will likely be nearly symmetric since there are nearly equal numbers of the two groups of players mentioned above.
c) The distribution will have wide variation since field goal kickers tend to be small while linemen frequently weigh over 300 pounds.

15 a) There will be two modes since SUVs will be heavier than compacts.
b) It is symmetric due to the equal number of cars in each group.
c) It has wide variation due to the differences between the groups.

17 a) The answer to this depends on the location of the bus stop, but for many stops, the distribution will have one mode with a few times during heavy use when times are short and a few times during very light use when times are long.
b) The distribution will be nearly symmetric for the reasons above.
c) The distribution is likely to have moderate to wide variation depending on the location of the bus stop.

19 a) The distribution will have one mode somewhere near the speed limit.
b) It will be right-skewed since there will be a few people who exceed the speed limit, but even fewer who are much below the limit.
c) There will be moderate variation, primarily due to the speeders.

21 a) The distribution is likely to have one mode.
b) The distribution is likely to be right-skewed. There will be a greater percentage of young people and families with children.
c) It will have wide variation since all ages visit amusement parks.

23 a) Since there will always be exactly 4 players, there is one mode.

b) The distribution is likely to be symmetric.
c) The distribution is likely to have no variation at all.

25 a) The distribution is likely to have one mode.
b) The distribution is likely to be symmetric.
c) The distribution is likely to have moderate variation.

27 a) The distribution is likely to have one mode since this is similar to the income distribution of all adults.
b) The distribution is likely to be right-skewed.
c) It will have wide variation, wider than in Exercise 22.

29 a) The distribution is likely to have one mode.
b) The distribution is likely to be right-skewed since low average players tend to not make the team.
c) The distribution is likely to have moderate variation.

Section 4.3

1 This statement is not sensible. A median score would be at the 50th percentile.

3 This statement is not sensible. There is less variation in the weights of ten hawks than ten birds of different species; therefore the boxplot for ten hawks should be narrower than the boxplot for ten birds of different species.

5 This statement is sensible. One would expect more variation in the gas mileages of ten different cars than in the gas mileages of ten identical cars.

7 $$\text{Mean} = \frac{6.2+9.3+4.1+5.2+6.7+7.7+11.0+7.2+7.7+5.6+8.5}{11} = \frac{79.2}{11} = 7.2$$

For the median, put the data in increasing order. Since there is an odd number of data values, the median is the middle (6th) value, or 7.2.
4.1 5.2 5.6 6.2 6.7 <u>7.2</u> 7.7 7.7 8.5 9.3 11.0

9 a) 835/1000 = 0.835, so you are in the 83rd percentile.
b) 921/1000 = 0.921, so you are in the 92nd percentile.
c) 125/1000 = 0.125, so you are in the 12th percentile.

11 a) 30% of 200 is 60, so the 30th percentile is around 60. While it is true that at least 30% (actually 30% exactly) of the values are less than or equal to 60 and at least 70% of the values (actually 70.5%) are greater than or equal to 60, the same is true for the number 61. At least 30% (actually 30.5%) of the values are less than or equal to 61 and at least 70% (actually 70%) of the values are greater than or equal to 61. Since 61 is between the minimum values for the 30th and 31st percentiles, we will use 61 as the 30th percentile.
b) 60% of 200 is 120. Following the same logic as in part a, the 60th percentile is 121.
c) 83% of 200 is 166. Following the same logic as in part a, the 83rd percentile is 167.

13 a) Set 1

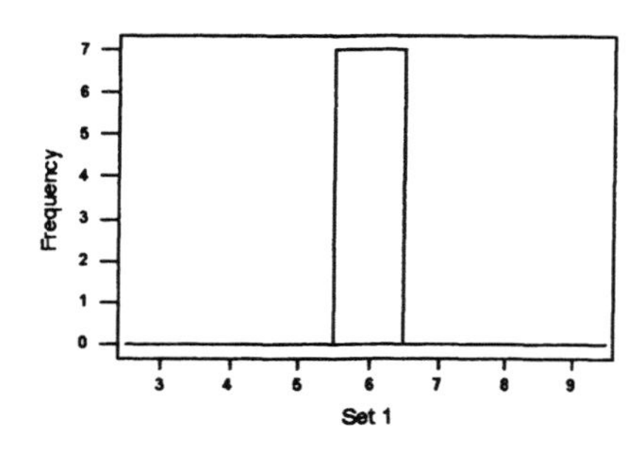

Set 2

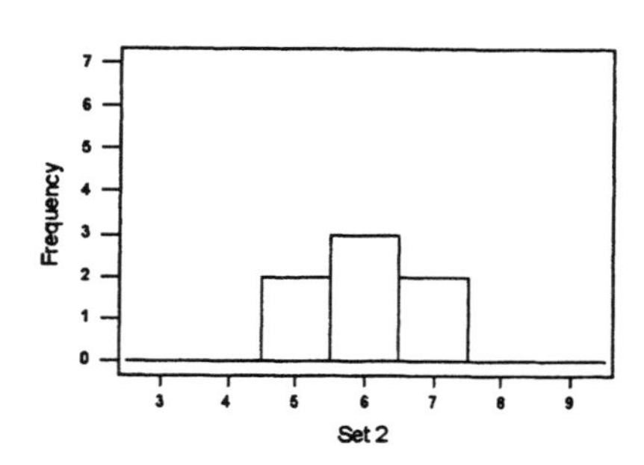

Set 3

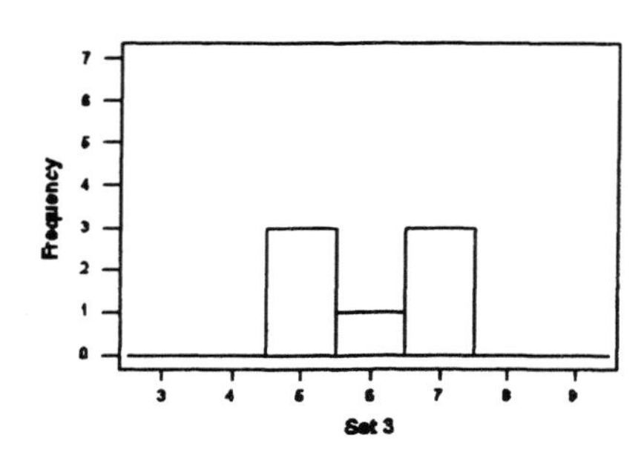

Set 4

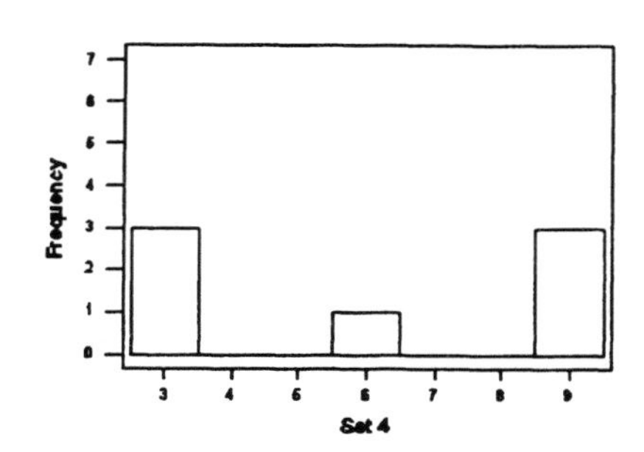

b) In each set, the median is the 4th value in the ordered list, the lower quartile is the middle value of the lowest three values (2nd in the overall list), and the upper quartile is the middle value of the highest three values (6th in the overall list).

	Set 1	Set 2	Set 3	Set 4
Low value	6	5	5	3
Lower quartile	6	5	5	3
Median	6	6	6	6
Upper quartile	6	7	7	9
High value	6	7	7	9

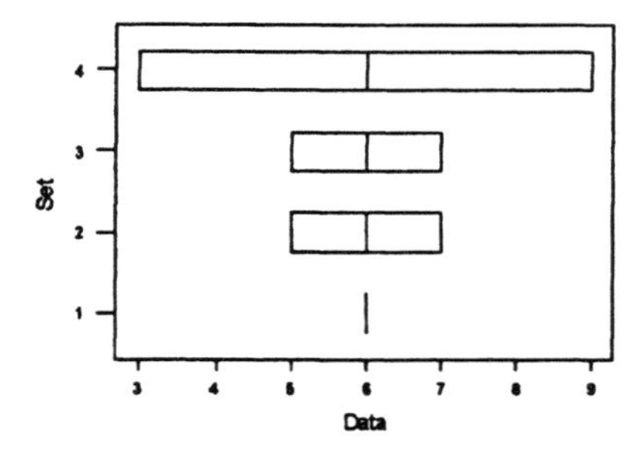

c) Set 1

Value	Deviation = Value - Mean	Deviation²
6	0	0
6	0	0
6	0	0
6	0	0
6	0	0
6	0	0
6	0	0
		Sum = 0

$$\text{Standard Deviation} = \sqrt{\frac{\text{Sum}}{7-1}} = \sqrt{\frac{0}{6}} = 0$$

Set 2

Value	Deviation = Value - Mean	Deviation²
5	-1	1
5	-1	1
6	0	0
6	0	0
6	0	0
7	1	1
7	1	1
		Sum = 4

$$\text{Standard Deviation} = \sqrt{\frac{\text{Sum}}{7-1}} = \sqrt{\frac{4}{6}} = 0.816$$

Set 3

Value	Deviation = Value - Mean	Deviation²
5	-1	1
5	-1	1
5	-1	1
6	0	0
7	1	1
7	1	1
7	1	1
		Sum = 6

Standard Deviation $= \sqrt{\frac{\text{Sum}}{7-1}} = \sqrt{\frac{6}{6}} = 1$

Set 4

Value	Deviation = Value - Mean	Deviation^2
3	-3	9
3	-3	9
3	-3	9
6	0	0
9	3	9
9	3	9
9	3	9
		Sum = 54

Standard Deviation $= \sqrt{\frac{\text{Sum}}{7-1}} = \sqrt{\frac{54}{6}} = 3$

d) The standard deviation takes all of the data into account and increases as the data become more spread out around the mean.

15 a) For the faculty, $\text{Mean} = \frac{2+3+1+0+1+2+4+3+3+2+1}{11} = \frac{22}{11} = 2$ years.

Median equals the sixth number in the ordered list and is 2 years.
Range = 4 - 0 = 4 years

For the students,

$\text{Mean} = \frac{5+6+8+2+7+10+1+4+6+10+9}{11} = \frac{68}{11} = 6.18$ years.

Median equals the sixth number in the ordered list and is 6 years.
Range = 10 - 1 = 9 years

b)

	Faculty	Students
Low Value	0	1
Lower quartile	1	4
Median	2	6
Upper quartile	3	9
High Value	4	10

The lower quartile is the middle value of the lowest 5 values in each data set and the upper quartile is the middle value of the highest 5 values.

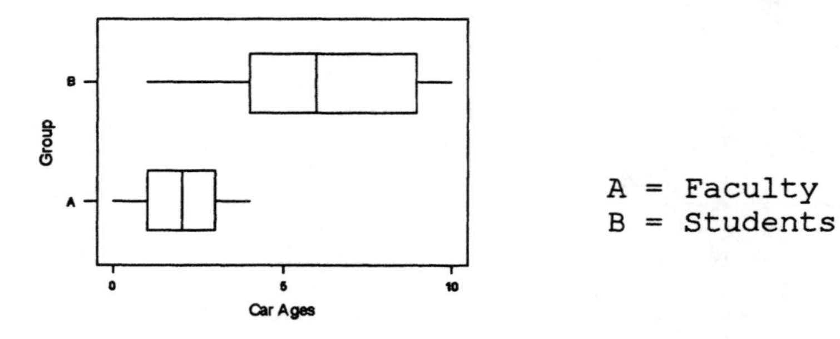

A = Faculty
B = Students

c) Computation of standard deviation

	Faculty			Students	
Value	**Deviation = Value - Mean**	**Deviation²**	**Value**	**Deviation = Value - Mean**	**Deviation²**
2	0	0	5	-1.18	1.3924
3	1	1	6	-.18	.0324
1	-1	1	8	1.82	3.3124
0	-2	4	2	-4.18	17.4724
1	-1	1	7	.82	.6724
2	0	0	10	3.82	14.5924
4	2	4	1	-5.18	26.8324
3	1	1	4	-2.18	4.7524
3	1	1	6	-.18	.0324
2	0	0	10	3.82	14.5924
1	-1	1	9	2.82	7.9524
2		**14**	**6.18**		**91.6364**

The means for faculty and students are given in bold at the bottoms of the first and fourth columns respectively. The deviations for faculty are obtained by subtracting the mean from each number in the first column. Similarly for students. The squared deviations are then placed in the third and sixth columns, and their totals are shown in bold at the bottom of the columns. The standard deviations are then found by dividing the sum of the squared deviations by n-1 = 11-1 = 10 and taking the square root. Thus, the standard deviations are

Faculty: $\text{Standard deviation} = \sqrt{\frac{14}{10}} = \sqrt{1.4} = 1.18$

Students: $\text{Standard deviation} = \sqrt{\frac{91.6364}{10}} = \sqrt{9.16364} = 3.03$

d) By the range rule, the standard deviation is approximately range/4, which for the faculty is 4/4 = 1 and for the students is 9/4 = 2.25. Both estimates are low, but are reasonably close.

e) The students have a higher mean age for their cars and a much greater variation in ages.

17 a) For the Stanley Cup, $\text{Mean} = \frac{4+5+7+4+4+4+4+6+6+7}{10} = \frac{51}{10} = 5.1$ games.

Median equals the average of the fifth and sixth numbers in the ordered list and is 4.5 games.
Range = 7 - 4 = 3 games

For the NBA, $\text{Mean} = \frac{6+6+7+4+6+6+6+5+6+5}{10} = \frac{57}{10} = 5.7$ games.

Median equals the average of the fifth and sixth numbers in the ordered list and is 6.0 games.
Range = 7 - 4 = 3 games

b)

	Stanley Cup	NBA
Low Value	4	4
Lower quartile	4	5
Median	4.5	6
Upper quartile	6	6
High Value	7	7

The lower quartile is the middle value of the lowest 5 values in each data set. The upper quartile is the middle value of the highest 5 values in each data set.

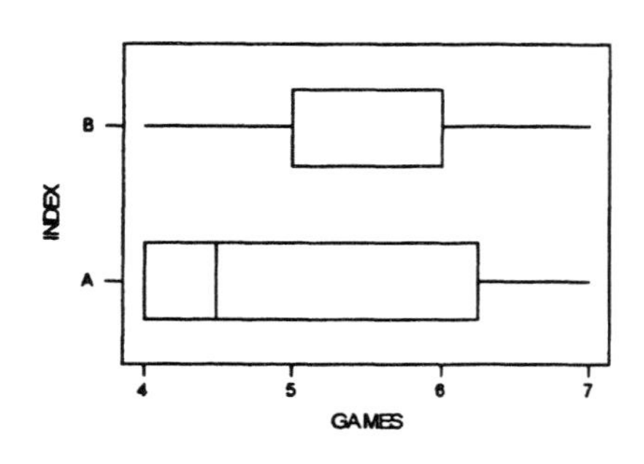

A = Stanley Cup
B = NBA

c) Computation of standard deviation

	Stanley Cup			**NBA**	
Value	**Deviation = Value - Mean**	**Deviation²**	**Value**	**Deviation = Value - Mean**	**Deviation²**
4	-1.1	1.21	6	.3	.09
5	-.1	.01	6	.3	.09
7	1.9	3.61	7	1.3	1.69
4	-1.1	1.21	4	-1.7	2.89
4	-1.1	1.21	6	.3	.09
4	-1.1	1.21	6	.3	.09
4	-1.1	1.21	6	.3	.09
6	.9	.81	5	-.7	.49
6	.9	.81	6	.3	.09
7	1.9	3.61	5	-.7	.49
5.1		**14.90**	**5.7**		**6.10**

The means for the Stanley Cup and the NBA are given in bold at the bottoms of the first and fourth columns respectively. The deviations for the Stanley Cup are obtained by subtracting the mean from each number in the first column. For the NBA, the deviations are obtained by subtracting the mean from each number in the fourth column. The squared deviations are then placed in the third and sixth columns, and their totals are shown in bold at the bottom of the columns. The standard deviations are then found by dividing the sum of the squared deviations by n-1 = 10-1 = 9 and taking the square root. Thus, the standard deviations are

Stanley Cup: $\text{Standard deviation} = \sqrt{\frac{14.9}{9}} = \sqrt{1.6556} = 1.29$

NBA: $\text{Standard deviation} = \sqrt{\frac{6.1}{9}} = \sqrt{.6778} = 0.82$

d) By the range rule, the standard deviation is approximately range/4, which for the Stanley Cup is 3/4 = 0.75 and for the NBA is 3/4 = 0.75. Both estimates are low, but are reasonably close.

e) The average series is shorter in the Stanley Cup, but the variation is slightly greater in the Stanley cup.

19 a) For Beethoven's symphonies,

$$\text{Mean} = \frac{28+36+50+33+30+40+38+26+68}{9} = \frac{349}{9} = 38.8 \text{ minutes.}$$

Median equals the fifth number in the ordered list and is 36

minutes.
Range = 68 - 26 = 42 minutes

For Mahler's symphonies,

$$\text{Mean} = \frac{52+85+94+50+72+72+80+90+80}{9} = \frac{675}{9} = 75.0 \text{ minutes.}$$

Median equals the fifth number in the ordered list and is 80 minutes.
Range = 94 - 50 = 44 minutes

b)

	Beethoven	Mahler
Low Value	26	50
Lower quartile	29	62
Median	36	80
Upper quartile	45	87.5
High Value	68	94

The lower quartile is the average of the two middle values in the lowest 4 values of the data set. The upper quartile is the average of the two middle values in the highest 4 values of the data set.

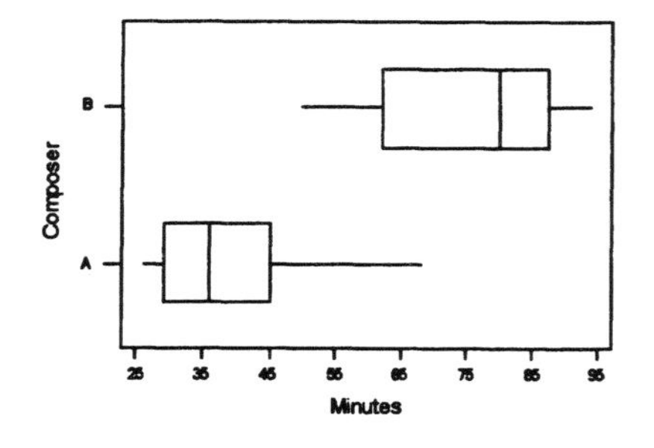

A = Beethoven's Symphonies
B = Mahler's Symphonies

c) Computation of standard deviation

	Beethoven			**Mahler**	
Value	**Deviation**	**Deviation^2**	**Value**	**Deviation**	**Deviation^2**
28	-10.8	116.64	52	-23	529
36	-2.8	7.84	85	10	100
50	11.2	125.44	94	19	361
33	-5.8	33.64	50	-25	625
30	-8.8	77.44	72	-3	9
40	1.2	1.44	72	-3	9
38	-.8	.64	80	5	25
26	-12.8	163.84	90	15	225
68	29.2	852.64	80	5	25
38.8		**1378.96**	**75.0**		**1908**

The mean lengths for Beethoven's and Mahler's symphonies are given in bold at the bottoms of the first and fourth columns respectively. The deviations for Beethoven's are obtained by subtracting the mean from each number in the first column. Similarly for Mahler's. The squared deviations are then placed in the third and sixth columns, and their totals are shown in bold at the bottom of the columns. The standard deviations are then found by dividing the sum of the squared deviations by n-1 = 9-1 = 8 and taking the square root. Thus, the standard deviations are

Beethoven: $$\text{Standard deviation} = \sqrt{\frac{1378.96}{8}} = \sqrt{172.37} = 13.13$$

Mahler: $$\text{Standard deviation} = \sqrt{\frac{1908}{8}} = \sqrt{238.5} = 15.44$$

d) By the range rule, the standard deviation is approximately range/4, which for Beethoven's symphonies is 42/4 = 10.5 and for Mahler's is 44/4 = 11.0. Both estimates are low, but are reasonably close.

e) The average length of Mahler's symphonies is much greater than that of Beethoven's, but the variation is about the same for both composers.

21 No. There are no negative percentiles.

23 The second shop has a slightly lower average delivery time, but its standard deviation is so large that you risk the pizza being delivered 20 to 40 minutes late. It could, of course, arrive much earlier than you expected as well. If you need to know the arrival time quite closely, you should order from the first shop, particularly since the average delivery time is only three minutes longer.

25 A lower standard deviation means more certainty in the value of the portfolio and less risk.

27 The batting averages are more closely bunched today than in the past. Since the overall average has remained at 0.260, averages above 0.350, should be less common today.

Chapter Review Exercises

1 a) Coke: $$\text{Mean} = \frac{0.8192 + 0.8150 + 0.8163 + \ldots + 0.8073 + 0.8079}{16} = \frac{13.0428}{16} = 0.81518$$

The median is the average of the eighth and ninth numbers (underlined) in the ordered list of 16 data values. Since the list is
0.7901 0.8062 0.8073 0.8079 0.8110 0.8128 0.8150 <u>0.8163</u> <u>0.8172</u> 0.8181 0.8192 0.8211 0.8244 0.8247 0.8251 0.8264, the median is (0.8163 + 0.8172)/2 = 0.81675.

Range = 0.8264 - 0.7901 = 0.0363

Diet Coke: $$\text{Mean} = \frac{0.7773 + 0.7758 + 0.7896 + \ldots + 0.7872 + 0.7813}{16} = \frac{12.5534}{16} = 0.78459$$

The median is the average of the eighth and ninth numbers

(underlined) in the ordered list of 16 data values. Since the list is
0.7758 0.7773 0.7806 0.7813 0.7823 0.7826 0.7830 0.7844 0.7852 0.7852 0.7861 0.7868 0.7872 0.7879 0.7881 0.7896, the Median is (0.7844 + 0.7852)/2 = 0.7848.

Range = 0.7896 - 0.7758 = 0.0038

b)

	Coke	Diet Coke
Low value	0.7901	0.7758
Lower quartile	0.80945	0.7818
Median	0.81675	0.7848
Upper quartile	0.82275	0.7870
High value	0.8264	0.7896

The lower quartiles are the averages of the 4th and 5th numbers in the ordered lists. The upper quartiles use the 12th and 13th numbers.

A = Regular Coke
B = Diet Coke

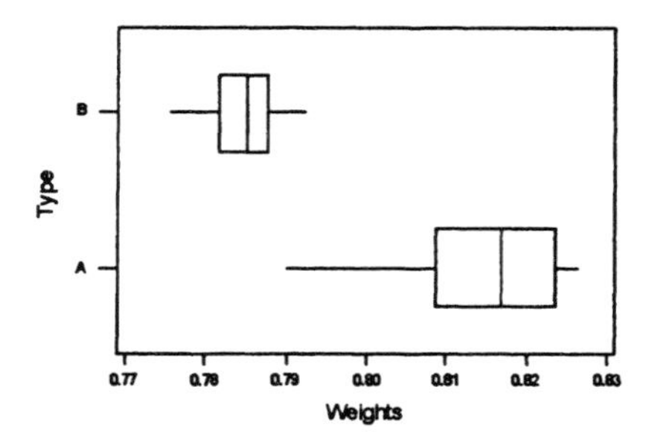

c)

Regular Coke	Deviations	Deviations^2	Diet Coke	Deviations	Deviations^2
.8192	.004025	.0000162	.7773	-	.0000531
.8150	-.000175	.0000000	.7758	-	.0000772
.8163	.001125	.0000013	.7896		.0000251
.8211	.005925	.0000351	.7868		.0000049
.8181	.002925	.0000086	.7844	-	.0000000
.8247	.009525	.0000907	.7861		.0000023
.8062	-.008975	.0000806	.7806	-	.0000159
.8128	-.002375	.0000056	.7830	-	.0000025
.8172	.002025	.0000041	.7852		.0000004
.8110	-.004175	.0000174	.7879		.0000110
.8251	.009925	.0000985	.7881		.0000123
.8264	.011225	.0001260	.7826	-	.0000040
.7901	-.025075	.0006288	.7923		.0000595
.8244	.009225	.0000851	.7852		.0000004
.8073	-.007875	.0000620	.7872		.0000068
.8079	-.007275	.0000529	.7813	-	.0000108
.81518		**.0013129**	**.78459**		**.0002862**

The mean weights for Regular Coke and Diet Coke are given in bold at the bottoms of the first and fourth columns respectively. The deviations for Regular Coke are obtained by subtracting the mean from each number in the first column. For Diet Coke, the deviations are obtained by subtracting the mean from each number in the fourth column. The squared deviations are then placed in the third and sixth columns, and their totals are shown in bold at the bottom of the columns. The standard deviations are then found by dividing the sum of the squared deviations by n-1 = 16-1 = 15 and taking the square root. Thus, the standard deviations are

Regular Coke: $\text{Standard deviation} = \sqrt{\dfrac{0.0013129}{15}} = \sqrt{0.0000875} = 0.0094$

Diet Coke: $\text{Standard deviation} = \sqrt{\dfrac{0.0002862}{15}} = \sqrt{0.00001908} = 0.0044$

d) By the range rule, the standard deviation is approximately range/4, which for Regular Coke is 0.0363/4 = 0.009075 and for Diet Coke is 0.0165/4 = 0.004125. Both estimates are low, but are reasonably close.

e) The Diet Coke weighs, on the average, about 0.03 pounds less per can and has about half the variation in weights that the Regular Coke cans have.

f) Cans of Diet Coke weigh less than cans of Regular Coke. Assuming that the cans are all the same size, it is likely that one of the ingredients in Regular Coke, perhaps the sugar, makes the liquid in Regular Coke slightly more dense than the liquid in Diet Coke.

2 a) Zero

b) This is a toss-up. While both batteries are equally likely to achieve a life length of 48 months, the batteries with a standard deviation of 2 months are likely to come closer to lasting exactly 48 months. Some of the batteries with a 6 month standard deviation will likely fail well before the 48 months is up, but an equal number of them will last somewhat beyond the 48 month period.

c) The outlier pulls the mean either up or down, depending on whether it is above or below the mean, respectively.

d) The outlier has no effect on the median since the median is found as the average of the two middle values in the ordered list (25th and 26th in a sample of size 50). The outlier would be either first or last in the list.

e) The outlier increases the range since one of the two numbers used to compute the range will be the outlier.

f) The outlier increases the standard deviation since one of the squared deviations will be larger than if it would be if there were no outlier.

CHAPTER 5 ANSWERS

Section 5.1

1 This statement is not sensible. A normal distribution is symmetric, so the mean and the median are equal.

3 This statement is not sensible. According to the discussion in the text, approximately 20% of all babies are two or more weeks overdue, so 40% of all babies cannot be three or more weeks overdue.

5 This statement is sensible. The errors in cutting and measuring 8-foot beams are random and hence approximately normally distributed.

7 Distribution b is not normal since it is not symmetric. Distribution c has the larger standard deviation since it is more spread out than distribution a.

9 Verbal SAT scores are normally distributed because test scores are the result of many different factors.

11 Candy bar weights are normally distributed because there will be small random variations in weight both above and below the mean weight. Note that the mean weight might be slightly above 4 ounces to ensure that almost all candy bars meet the advertised weight.

13 Travel times will be normally distributed because there will be random variations in travel time both above and below the mean time.

15 Potato chip bag weights are normally distributed because there will be small random variations in weight both above and below the mean weight. Note that the mean weight might be slightly above the advertised weight to ensure that almost all candy bars meet the advertised weight.

17 If prices at the dealers are independent of each other, they would be nearly normally distributed. Price wars or price fixing could lead to other distributions.

19 The movie lengths are not normally distributed because movies generally have a minimum length, but no maximum length. There are very few movies that are a lot shorter than a typical length, but there are some very long movies.

21 The quarter weights should be nearly normal because the deviations from the mean are symmetrical above and below the mean.

23 a) The distributions that are closest to normal are median age (b) and incarceration rate [c]. Population density [a] is very skewed to the right.
   b) The answer to this question will vary depending on which of the variables in Appendix A is chosen. For example, "%Urban Land" and "%Foreign Born" are clearly not normally distributed.

25 a) The total area under the curve is 1.
   b) 0.50
   c) 0.30

d) 0.70
e) 0.20

27 a) The mean is 155.
b) 20%
c) 20%
d) 45%

29 a) About 700 of the 5738 men had chest sizes less than 38 inches. This is 12%.
b) About 180 of the 5738 men had chest sizes greater than 43 inches. This is a relative frequency of 0.03.
c) 88% (100% - 12% from part (a))
d) About 2530 of the 5738 men had chest sizes less than or equal to 39 inches. This is a relative frequency of 0.44.

Section 5.2

1 This statement is not sensible. If we assume that the minimum test score is zero, then the standard deviation cannot be greater than the mean.

3 This statement is sensible, if we assume that the scores are normally distributed. Approximately 64% of the values in a normal distribution are less than one standard deviation above the mean.

5 This statement is not sensible. A negative *z*-score places a data value below the mean. A score one standard deviation above the mean has a positive *z*-score.

7 a) 50%. Half of all values are less than the mean.
b) 0.84. 120 is 1 standard deviation to the right of the mean. There is .50 to the left of 100 and another .34 between 100 and 120.
c) 97.5%. 140 is 2 standard deviations to the right of the mean. There is .50 to the left of 100 and another .475 between 100 and 140.
d) 16%. 80 is 1 standard deviation to the left of the mean. The percent of the scores between 80 and the mean is 34%, so there is 16% to the left of 80.
e) 0.025. 60 is two standard deviations to the left of the mean. The relative frequency of scores between 60 and the mean is .475, so there is 0.025 to the left of 60.
f) 16%. 120 is 1 standard deviation to the right of the mean. The percent of the scores between 120 and the mean is 34%, so there is 16% to the right of 120.
g) 2.5%. 140 is two standard deviations to the right of the mean. The percentage of scores between 140 and the mean is 47.5%, so there is 2.5% to the right of 140.
h) 0.84. From part (d), 16% of the scores are less than 80, so 84% must be greater than 80.
i) 68%. 80 and 120 are both 1 standard deviation from the mean.
j) 81.5%. 80 is 1 standard deviation below the mean and 140 is 2 standard deviations above the mean. There is 34% between 80 and 100 and 47.5% between 100 and 140, a total of 81.5%.

9 In all cases, 5% of coins are rejected. The ranges of weights that are acceptable to the vending machine are

Cent	2.44-2.56 grams
Nickel	4.88-5.12 grams
Dime	2.208-2.328 grams
Quarter	5.53-5.81 grams
Half dollar	11.06-11.62 grams

11 a)

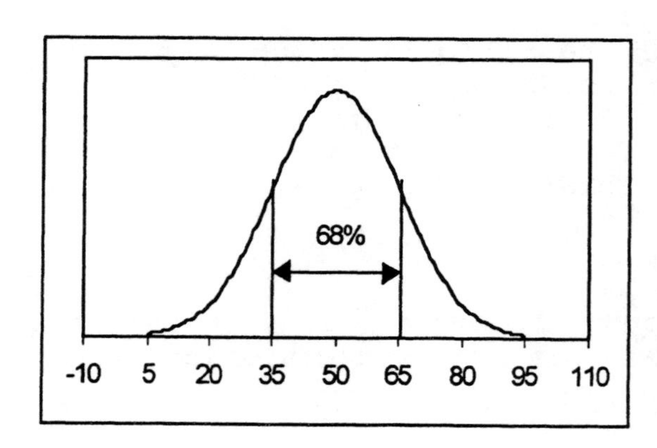

b)

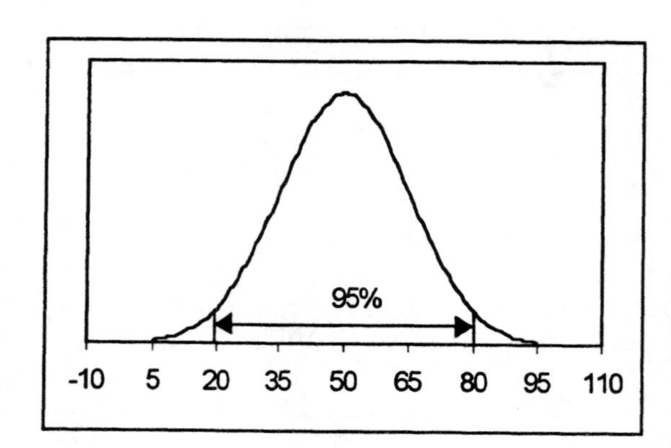

13
a) The standard score is 0.5 which is at the 69.15 percentile.
b) The standard score is -1 which is at the 15.87 percentile.
c) The standard score is 2 which is at the 97.72 percentile.
d) The standard score is 1.9 which is at the 97.13 percentile.

15
a) 1.0 standard deviations above the mean.
b) 1.3 standard deviations below the mean.
c) 0.1 standard deviations above the mean.
d) 0.75 standard deviations below the mean.

17
a) $\text{Standard score} = \dfrac{\text{data value - mean}}{\text{standard deviation}} = \dfrac{260-268}{15} = -0.53$; approximately 30 percentile

b) $\text{Standard score} = \dfrac{\text{data value - mean}}{\text{standard deviation}} = \dfrac{270-268}{15} = 0.13$; approximately 56 percentile

c) Standard score $= \dfrac{\text{data value - mean}}{\text{standard deviation}} = \dfrac{255-268}{15} = -0.87$; approximately 19 percentile

d) Standard score $= \dfrac{\text{data value - mean}}{\text{standard deviation}} = \dfrac{265-268}{15} = -0.20$; approximately 42 percentile

19 a) Standard score $= \dfrac{\text{data value - mean}}{\text{standard deviation}} = \dfrac{650-497}{115} = 1.33$; approximately 90.8 percentile, interpolating between the percentiles for standard scores of 1.30 and 1.40.

b) Using Table 5.1, the standard score required is approximately 1.65, so the GRE score will be 1.65 standard deviations above the mean, or 1.65 x 115 = 190 above the mean. Thus the GRE score required is 497 + 190 = 687.

21 Standard score $= \dfrac{\text{data value - mean}}{\text{standard deviation}} = \dfrac{33000-30000}{6000} = 0.50$; the standard score for 27000 is -0.50. The percentiles for 33000 and 27000 are therefore, from Table 5.1, 69.15 and 30.85; i.e., 69.15% earn less than $33000, but 30.85% earn less than $27000. This means that 69.15% - 30.85% = 38.30% of the workers earn between $27000 and $33000.

23 a) Standard score $= \dfrac{\text{data value - mean}}{\text{standard deviation}} = \dfrac{31-30.4}{0.23} = 2.61$; this is at the 99.53 percentile. Thus 99.53% of the barometers read <u>below</u> 31, so 0.47% read above 31. Since only 50 barometers are being read, and 0.0047 x 50 = 0.235, there are probably no barometers reading above 31 out of the 50 tested.

b) Standard score $= \dfrac{\text{data value - mean}}{\text{standard deviation}} = \dfrac{30-30.4}{0.23} = -1.74$; this is approximately at the 4th percentile, so approximately 4% of the barometers read less than 30.

c) The company will reject barometers that read 1.5 standard deviations or more, or 1.5 x .23 = 0.345, above or below the mean. Thus the lower critical reading is 30.400 - 0.345 = 30.055 inches and the upper critical reading is 30.400 + 0.345 = 30.745 inches.

d) The best estimate would be the mean of the readings, or 30.4 inches, since the mean is more reliable than any single reading.

25 Standard score $= \dfrac{\text{data value - mean}}{\text{standard deviation}} = \dfrac{64-69}{2.8} = -1.79$

Standard score $= \dfrac{\text{data value - mean}}{\text{standard deviation}} = \dfrac{78-69}{2.8} = 3.21$

78 inches has a standard score of 3.21 which is approximately at the 99.91 percentile; 64 inches has a standard score of -1.79 which is approximately at the 3.58 percentile; thus approximately 99.91 - 3.58 = 96.33 percent of American men are eligible for the Marines.

27 a) By the 68-95-99.7 rule, 68% of the data values are within one standard deviation of the mean.

b) One standard deviation either side of the mean includes the interval 39.85 ± 2.07 or 37.78 to 41.92. This includes the chest sizes listed in the table as 38, 39, 40, and 41. The number of such data values is 749 + 1073 + 1079 + 934 = 3835. Thus 3835/5738 = 0.668, or 66.8% of the chest sizes fall within one standard deviation of the mean.

c) By the 68-95-99.7 rule, 95% of the data values are within two standard deviations of the mean.

d) Two standard deviations either side of the mean includes the interval 39.85 ± 2(2.07) or 35.77 to 43.99. This includes the chest sizes listed in the table as 36, 37, 38, 39, 40, 41, 42, and 43. The number of such data values is 185 + 420 + 749 + 1073 + 1079 + 934 + 658 + 370 = 5468. Thus 5468/5738 = 0.953, or 95.3% of the chest sizes fall within two standard deviations of the mean.

e) This distribution is very close to a normal distribution.

Section 5.3

1 The mean of the distribution of sample means is the population mean, 100. With n = 100, the standard deviation of the distribution of means is $\sigma/\sqrt{n} = 16/\sqrt{100} = 1.6$; with n = 400, the standard deviation is $\sigma/\sqrt{n} = 16/\sqrt{400} = 0.8$. With larger sample sizes, the distribution of sample means is narrower (smaller standard deviation) due to the higher reliability of the larger samples.

3 The mean of the distribution of sample means is the population mean, 6.5. With n = 36, the standard deviation of the distribution of means is $\sigma/\sqrt{\text{n}} = 3.452/\sqrt{36} = 0.58$; with n = 100, the standard deviation is $\sigma/\sqrt{\text{n}} = 3.452/\sqrt{100} = 0.35$. The distribution of sample means should be near normal with n = 36 and closer to normal with n = 100 since as n increases, the distribution becomes more normal .

5 a) The standard deviation of the mean is $\sigma/\sqrt{n} = 2.07/\sqrt{25} = 0.414$; for a value of 40, the $\text{standard score} = \dfrac{\text{data value - mean}}{\text{standard deviation}} = \dfrac{40-39.85}{0.414} = 0.36$. From Table 5.1, this corresponds to about the 64th percentile. Thus, the likelihood that the sample mean will be greater than 40 is about 1.00 - .64 = 0.36.

b) The standard deviation of the mean is $\sigma/\sqrt{n} = 2.07/\sqrt{100} = 0.207$; for a value of 39.5, the $\text{standard score} = \dfrac{\text{data value - mean}}{\text{standard deviation}} = \dfrac{39.5-39.85}{0.207} = -1.69$. From Table 5.1, this corresponds to about the 4.4 percentile. Thus, the likelihood that the sample mean will be less than 39.5 is about 0.044.

c) The standard deviation of the mean is $\sigma/\sqrt{n} = 2.07/\sqrt{625} = 0.0828$; for a value of 40, the $\text{standard score} = \dfrac{\text{data value - mean}}{\text{standard deviation}} = \dfrac{40-39.85}{0.0828} = 1.81$. From Table 5.1, this corresponds to about the 96.4 percentile. Thus, the likelihood that the sample mean will be greater than 40 is about 1.000 - 0.964 = 0.036.

d) The standard deviation of the mean is $\sigma/\sqrt{n} = 2.07/\sqrt{1600} = 0.05175$; for

a value of 39.7, the $\text{standard score} = \dfrac{\text{data value - mean}}{\text{standard deviation}} = \dfrac{39.7-39.85}{0.05175} = -2.90$.
From Table 5.1, this corresponds to about the 0.19 percentile. Thus, the likelihood that the sample mean will be less than 39.7 is about 0.0019.

7 $\text{Standard score} = \dfrac{\text{data value - mean}}{\text{standard deviation}} = \dfrac{15-13.0}{7.9} = 0.25$; this corresponds to the 60th percentile, so 40% of the individual aircraft have ages greater than 15 years. For a sample of 49 aircraft, the standard deviation of the mean is $\sigma/\sqrt{n} = 7.9/\sqrt{49} = 7.9/7 = 1.13$; thus
$\text{standard score} = \dfrac{\text{data value - mean}}{\text{standard deviation}} = \dfrac{15-13.0}{1.13} = 1.77$; this corresponds to the 96th percentile, so 4% of the sample mean ages are greater than 15 years.

9 $\text{Standard score} = \dfrac{\text{data value - mean}}{\text{standard deviation}} = \dfrac{10-13.0}{7.9} = -0.38$; interpolating in Table 5.1, this corresponds to the 35.4th percentile.
$\text{Standard score} = \dfrac{\text{data value - mean}}{\text{standard deviation}} = \dfrac{16-13.0}{7.9} = 0.38$; interpolating in Table 5.1, this corresponds to the 64.6th percentile. Thus 64.8%-35.4% = 29.4% of the individual aircraft have ages between 10 and 16 years.

For a sample of size n = 81, the standard deviation of the distribution of means is $\sigma/\sqrt{n} = 7.9/\sqrt{81} = 7.9/9 = 0.88$. For 10 years, the
$\text{Standard score} = \dfrac{\text{data value - mean}}{\text{standard deviation}} = \dfrac{10-13.0}{.88} = -3.41$; the standard score for 16 years is +3.41. These standard scores correspond to the 0.02 and 99.98 percentiles, so 99.98% - 0.02% = 99.96% of the means will lie between 10 and 16.

11 $\text{Standard score} = \dfrac{\text{data value - mean}}{\text{standard deviation}} = \dfrac{65-61}{9} = 0.44$; this corresponds to the 67th percentile, so 33% of the finishing times are greater than 61 minutes. For a sample of 25 runners, the standard deviation of the mean is
$\sigma/\sqrt{n} = 9/\sqrt{25} = 9/5 = 1.8$; thus the $\text{standard score} = \dfrac{\text{data value - mean}}{\text{standard deviation}} = \dfrac{65-61}{1.8} = 2.22$;
this corresponds to the 98.6 percentile, so 1.4% of the sample mean finishing times are greater than 61 minutes.

13 For a time of 59 minutes, the $\text{standard score} = \dfrac{\text{data value - mean}}{\text{standard deviation}} = \dfrac{59-61}{9} = -0.22$;
interpolating in Table 5.1, this corresponds to the 42nd percentile. For a time of 62 minutes, the $\text{standard score} = \dfrac{\text{data value - mean}}{\text{standard deviation}} = \dfrac{62-61}{9} = 0.11$;
interpolating in Table 5.1, this corresponds to the 54th percentile. Thus 54%-42% = 12% of the finishing times are between 59 and 61 minutes.

For a sample of size n = 64, the standard deviation of the distribution of means is $\sigma/\sqrt{n} = 9/\sqrt{64} = 9/8 = 1.125$. For 59 minutes, the

$\text{standard score} = \dfrac{\text{data value - mean}}{\text{standard deviation}} = \dfrac{59-61}{1.125} = -1.78$; similarly, the standard score for 62 minutes is 0.89. These standard scores correspond to the 4th and 81st percentiles, so 81% - 4% = 77% of the sample mean finishing times will lie between 59 and 61 minutes.

15 For a sample of size n = 100, the standard deviation of the distribution of means is $\sigma/\sqrt{n} = 0.290/\sqrt{100} = 0.029$.

a) For 0.55, the $\text{standard score} = \dfrac{\text{data value - mean}}{\text{standard deviation}} = \dfrac{0.55-0.50}{0.029} = 1.72$; this standard score corresponds to the 95.5 percentile, so 100% - 95.5% = 4.5% of the sample means will be greater than 0.55.

b) For 0.48, the $\text{standard score} = \dfrac{\text{data value - mean}}{\text{standard deviation}} = \dfrac{0.48-0.50}{0.029} = -0.69$; this standard score corresponds to the 24th percentile, so 24% of the sample means will be less than 0.48.

b) From part (a), 0.55 has a standard score of 1.72; similarly, 0.45 has a standard score of -1.72. The percentiles for these standard scores are 95.7 and 4.3, respectively. Thus 95.7% - 4.3% = 91.4% of the means will lie between 0.45 and 0.55.

c) From part (b), 0.48 has a standard score of -0.69; for 0.53, the $\text{standard score} = \dfrac{\text{data value - mean}}{\text{standard deviation}} = \dfrac{0.53-0.50}{0.029} = 1.03$. These standard scores correspond to the 24th and 84th percentiles, so 84% - 24% = 60% of the sample means will lie between 0.48 and 0.53.

Chapter Review Exercises

1

a) The numbers will occur with equal likelihood. Therefore, they have a uniform distribution, not a normal distribution.

b) Incomes are not generally normally distributed; they tend to be right-skewed and have no upper limit.

c) Test scores are often normally distributed since they are the result of numerous factors.

2

a) $\text{Standard score} = \dfrac{\text{data value - mean}}{\text{standard deviation}} = \dfrac{99.0-98.2}{0.62} = 1.29$; this corresponds to the 90th percentile.

b) From part (a), the standard score is 1.29.

c) No, the data value lies less than 2 standard deviations from the mean.

d) For a sample of size n = 50, the standard deviation of the distribution of means is $\sigma/\sqrt{n} = 0.62/\sqrt{50} = 0.0877$. For 97.98, the $\text{standard score} = \dfrac{\text{data value - mean}}{\text{standard deviation}} = \dfrac{97.98-98.20}{0.0877} = -2.51$. This corresponds to the 0.65 percentile, so the likelihood that the mean body temperature is 97.98 degrees is 0.0065.

e) For 101.00, the $\text{standard score} = \dfrac{\text{data value - mean}}{\text{standard deviation}} = \dfrac{101-98.20}{0.62} = 4.52$. This is an unusual temperature since it lies more than two standard deviations above the mean. We conclude that the person has a fever.

f) The 95th percentile is associated with a standard score of 1.65. Thus the temperature must be 1.65 standard deviations above the mean, or 1.65 x 0.62 = 1.02 degrees above 98.2 degrees, or 99.22.

g) The 5th percentile is associated with a standard score of -1.65. Thus the temperature must be 1.65 standard deviations below the mean, or 1.02 degrees below 98.2 degrees, or 97.18.

h) For 100.6, the $\text{standard score} = \dfrac{\text{data value - mean}}{\text{standard deviation}} = \dfrac{100.60 - 98.20}{0.62} = 3.87$. This corresponds to a percentile higher than 99.98. Thus fewer than 0.02% of healthy adults would be expected to have a temperature above 100.6. It should be very safe to conclude that someone with a temperature of 100.6 or higher has a fever. On the other hand, using such a high cutoff point will certainly result in concluding that a number of people do not have fevers when, in fact, they do have a fever.

i) For a sample of n = 106, the standard deviation of the distribution of sample means is $\sigma / \sqrt{n} = 0.62 / \sqrt{106} = 0.0602$. For a temperature of 98.2, the

$\text{standard score} = \dfrac{\text{data value - mean}}{\text{standard deviation}} = \dfrac{98.20 - 98.60}{0.0602} = -6.64$. Thus the sample mean is about 6.6 standard deviations below the mean; the chance of selecting such a sample is extremely small if the assumed mean is correct. The assumed mean (98.60) may be incorrect.

CHAPTER 6 ANSWERS

Section 6.1

1 This statement is not sensible. Instead of referring to a meaningful effect, statistical significance refers to an observed result that is unlikely to occur by chance.

3 The statement is sensible. Even though the exact probability is unknown, the difference is so extreme that it appears to be an observed event that is highly unlikely to occur by chance.

5 The statement is not sensible. Instead of easing headache pain for everyone, the difference could be significant by easing headache pain for a substantial number of people. That is, the treatment might be effective, but not 100% effective.

7 Significant. The coin is likely to be biased toward heads.

9 Not significant. We only expect about 1/6 of 100, or 17, threes.

11 Significant. It is likely that the team has improved considerably, possibly by means of a trade or the addition of new player(s).

13 Not significant. This is only one win less than expected (65% of 20 = 13).

15 Significant. The player would be expected to make only about 42 or 43 out of 50 and is obviously on a hot streak.

17 Significant. One would not expect a difference this great among thirty cars driven under identical conditions.

19 a) If 100 samples were selected, the mean temperature would be 98.20 or less in 5 or fewer of the samples.
b) Selecting a sample with a mean this small is extremely unlikely and would not be expected by chance.

21 This result is not significant because the probability of it occurring by chance when there is no real improvement is greater than 0.05.

Section 6.2

1 This statement is not sensible. If the statement were true, a consequence would be that approximately 50% of all planets have life, which is not the case. We might not have enough information to determine the probability of life on Pluto.

3 The statement is sensible because the probability of the complement is 1 - 0.3 = 0.7.

5 The statement is not sensible because the subjective probability is much higher than the likely correct value.

7 a) Outcome
b) Event since it can occur in three different ways

c) Outcome
d) Event since it can occur in three different ways
e) Event since it can occur in six different ways
f) Event since it can occur in six different ways

9 $$\text{P(Rolling an odd number)} = \frac{\text{Number of ways an odd number can occur}}{\text{Total number of outcomes}} = \frac{3}{6} = 0.5$$

We assume that all sides of the die are equally likely.

11 $$\text{P(Tuesday)} = \frac{\text{Number of ways a Tuesday can occur}}{\text{Total number of outcomes}} = \frac{1}{7} = 0.14$$

We assume that all days of the week are equally likely for a birth.

13 $$\text{P(0)} = \frac{\text{Number of ways 0 can occur}}{\text{Total number of outcomes}} = \frac{1}{10} = 0.1$$

We assume that all digits are equally likely for the last digit.

15 $$\text{P(Between midnight and 1:00AM)} = \frac{\text{Number of hours between 12 and 1}}{\text{Total Number of hours in a day}} = \frac{1}{24} = 0.04$$

We assume that all times of the day are equally likely for a birth.

17 $$\text{P(Three girls)} = \frac{\text{Number of ways three girls can occur}}{\text{Total number of outcomes}} = \frac{3}{8} = 0.375$$

We assume that GGG, BGG, GBG, GGB, BBG, BGB, GBB, and BBB are equally likely outcomes.

19 P(not rolling a 4) = 1 - P(rolling a 4) = 1 - 1/6 = 5/6 = 0.833

21 P(not tossing 2 heads) = 1 - P(tossing 2 heads) = 1 - 1/4 = 3/4 = 0.75

23 P(not January, February, or March) = 1 - P(January, February, or March) = 1 - 3/12 = 9/12 = 0.75. This answer is approximate and based on the assumption that all months are equally likely for a birth. This assumption is not exactly true, first because the months don't have equal numbers of days, and second, because births are not uniformly distributed across the calendar.

25 P(not 4 and not 5) = 1 - P(4 or 5) = 1 - 2/6 = 4/6 = 0.67

27 $$\text{P(Red M\&M)} = \frac{\text{Number of ways a Red M\&M can occur}}{\text{Total number of outcomes}} = \frac{10}{45} = 0.22$$

$$\text{P(Blue M\&M)} = \frac{\text{Number of ways a Blue M\&M can occur}}{\text{Total number of outcomes}} = \frac{15}{45} = 0.33$$

$$\text{P(Yellow M\&M)} = \frac{\text{Number of ways a Yellow M\&M can occur}}{\text{Total number of outcomes}} = \frac{20}{45} = 0.44$$

$$\text{P(Red or Blue M\&M)} = \frac{\text{Number of ways a Red or Blue M\&M can occur}}{\text{Total number of outcomes}} = \frac{25}{45} = 0.55$$

In all of these cases, we assume that each of the M&Ms is equally likely to be drawn.

29 You should expect to get about 20% since you have a 1/5 chance of getting each question correct. This assumes that you are equally likely to choose each answer.

31 There are eight possible outcomes: GGG, BGG, GBG, GGB, BBG, BGB, GBB, BBB
For each of the parts below, the solution is obtained by dividing the number of outcomes in the event by 8.

a) 1/8 = 0.125 (GGG)
b) 3/8 = 0.375 (BBG, BGB, GBB)
c) 1/8 = 0.125 (GBB)
d) 7/8 = 0.875 (GGG, BGG, GBG, GGB, BBG, BGB, GBB)
e) 4/8 = 0.5 (BBG, BGB, GBB, BBB)

33 12/30 = 0.40

35 0.01, since a 100-year flood occurs about once in 100 years.

37 306/281,000,000 = 0.00000109

39 The probability that a person you meet at random is over 65 will be 34.7 million/281 million = 0.123 in 2000 and will be 78.9/394 = 0.200 in 2050. Thus your chances will be greater in 2050.

41 P(not a man) = 1 - P(man) = 1 - (28 + 16 + 4)/100 = 1 - 48/100 = 0.52
P(not a Republican) = 1 - P(Republican) = 1 - (21 + 28)/100 = 1 - 49/100 = 51/100 = 0.51

43 a) 0.045 b) 0.066
c)

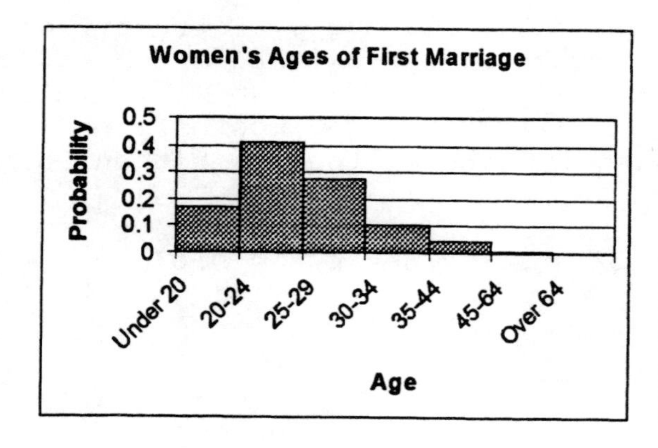

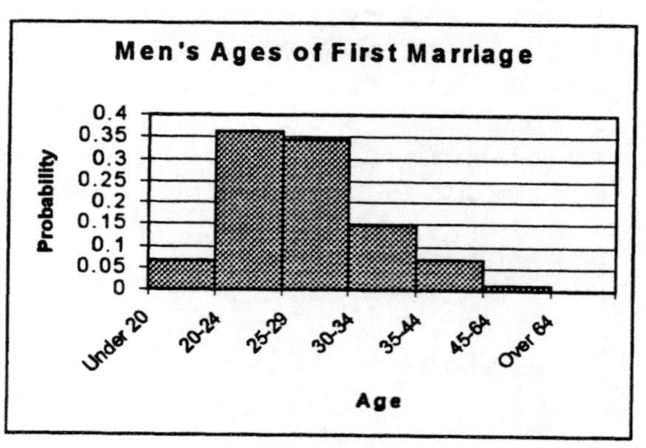

45 a) 1/42 = 0.024 since each of the numbers is equally likely
b) 163/5964 = 0.027 for a 34
c) 116/5964 = 0.019 for a 6
d) The deviations are what might be expected by chance.

Section 6.3

1 This statement is not sensible. According to the law of large numbers, even though Scott might start to win more than he has been winning, his losses will probably increase.

3 This statement is not sensible. The coin has no memory, so the outcome on the tenth toss is independent of the previous tosses. Assuming that the coin is fair, heads and tails would be equally likely on the tenth toss.

5 No, you should not expect to get exactly 5000 heads since the probability of that particular outcome is extremely small. The proportion of heads should approach 0.5 as the number of tosses increases.

7 Looking at Table 6.2, there are nine outcomes containing two even numbers. Thus the probability of getting two even numbers when you roll two fair dice is 9/36 = 0.25, and the probability of not getting two even numbers is 1 - 0.25 = 0.75. The expected value to you for one game is ($5 x 0.25) + (-$1 x 0.75) = $1.25 - $.75 = $0.50. Since you must either win $5 or lose $1 on the first game, you can not win the expected value of $0.50. Even if the expected value were equal to one of the outcomes, the outcome of one game can not be predicted, so you should not actually expect to win the expected value. If you play 100 games, winning an average of $0.50 per game, you should expect to win 100 x $0.50 = $50.

9 For the 1-point attempt, Expected value = (1 x 0.94) + (0 x 0.06) = 0.94
For the 2-point attempt, Expected value = (2 x 0.37) + (0 x 0.63) = 0.74
The 1-point attempt makes more sense in most cases. However, if a team is two points behind with little or no time left, it makes no sense to go for 1 point. The same is true if the team is behind by 16 points and it is unlikely that the team will have enough time to score three times.

11 Your waiting time is uniformly distributed between 0 and 30 minutes. The center of this symmetric distribution is 15 minutes, so 15 minutes is your expected waiting time.

13 On one ticket, you will spend one dollar and you may get something back. The expected value is -$1 + ($3,000,000 x 1/76,275,360) + ($150,000 x 1/2,179,296) + ($5,000 x 1/339,002) + ($150 x 1/9,686) + ($100 x 1/7,705) + ($5 x 1/220) + ($5 x 1/538) + ($2 x 1/102) + ($1 x 1/62) = -$0.78. If you spend $365 per year on the lottery, you can expect to lose 365 x $0.78 = $285.02.

15 Expected age = (7 x 0.20) + (19 x 0.153) + (29.5 x 0.136) + (39.5 x 0.163) + (54.5 x 0.222) + (N x 0.126) where N is chosen to be representative of the 65 and over group. If N is chosen to be 65, the expected age is about 35 years. If N is chosen to be 75, the expected age is 36.3 years. Since the average age is not very sensitive to fairly large changes in the age used to represent the oldest group, it is safe to say that the expected age is about 35 or 36.

17 a) **Decision 1** - Option A: Expected value = $1,000,000
Option B: Expected value = (%2,500,000 x 0.10) + ($1,000,000 x 0.89) + ($0 x 0.01) = $1,140,000

Decision 2 - Option A: Expected value = (%1,000,000 x 0.11) + ($0 x 0.89) = $110,000
Option B: Expected value = ($2,500,000 x 0.10) + ($0 x 0.90) = $250,000

b) Responses are not consistent with expected values in Decision 1, but they are consistent in Decision 2.

c) It appears that people choose the certain outcome ($1,000,000) in Decision 1.

19 a) 0.5; 0.5

b) You have gained $45 and lost $55, so you have a net loss of $10.

c) You have gained $92 and lost $108, so you have a net loss of $16.

d) You have gained $240 and lost $260, so you have a net loss of $20.

e) After 100 tosses, it was 45%; after 200 tosses, it was 92/200 or 46%; and after 500 tosses, it was 240/500 or 48%. This illustrates the gambler's fallacy because, even though the percentage of wins is increasing, the amount of the net loss is also increasing.

f) You would need to have 300 heads in 600 tosses. Since you already have 240 heads in the first 500 tosses, you would need 60 more heads in the last 100 tosses.

Section 6.4

1 This statement is not sensible. The lottery numbers are selected in a way that is independent of any previous results, so the outcomes of previous lottery selections have no effect on future results.

3 This statement is not sensible. If the probability of event *A* is 0.4, the probability of events *A* or *B* must be at least 0.4 and it cannot be 0.3.

5 Independent, since replacing the first card in the deck keeps the probability of an ace on the second card the same no matter what card was drawn the first time. The probability of getting two aces in a row is (4/52) x (4/52) = 1/169 = 0.0059.

7 The probability that a single phone number ends in a 1 is 1/10. Since your friends' phone numbers are independent of each other, the probability that all five end in a 1 is 1/10 x 1/10 x 1/10 x 1/10 x 1/10 = 1/100000 = 0.00001.

9 Assuming that the gender of each child is independent of that of the others, the probability of having 5 girls in a row is ½ x ½ x ½ x ½ x ½ = 1/32 = 0.031.

11 The births are independent of each other and each birth has P(boy) = 1/2, so P(four boys in a row) = ½ x ½ x ½ x ½ = 1/16 = 0.0625.

13 Since the selections can be repeated, the probabilities of each type remain the same for each selection: 30/60 = 1/2 for rock, 15/60 = 1/4 for jazz, and 15/60 = 1/4 for blues.

a) P(four jazz selections in a row) = 1/4 x 1/4 x 1/4 x 1/4 = 1/256 = 0.0039.

b) P(five blues in a row) = 1/4 x 1/4 x 1/4 x 1/4 x 1/4 = 1/1024 = 0.00098.

c) P(jazz and then rock) = 1/4 x 1/2 = 1/8 = 0.125.

d) Since P(non-rock selection) = 1 - P(rock) = 1 - 1/2 = 1/2, P(four non-rock selections in a row) = ½ x ½ x ½ x ½ = 1/16 = 0/.0625

e) There are 60 equally likely songs available for each selection. No matter which song is played first, the probability that the next one is the same is 1/60 = 0.0167.

15 The number of people who either pled guilty or were sent to prison is 392 + 564 + 58 = 1014. Therefore P(guilty plea or sent to prison) = 1014/1028 = 0.986.

17 a) P(B number) = 15/75 = 0.20

b) P(two B numbers in a row) = P(B number on first draw) x P (B number on second draw|B number on the first draw) = 15/75 x 14/74 = 0.038.

c) P(B number or O number) = 30/75 = 0.400.

d) P(B number, then an I number, then an N number) = P(B number) x P(I number|B number) x P(N number|B number and an I number) = 15/75 x 15/74 x 15/73 = 0.0083

e) P(five non-B numbers) = 60/75 x 59/74 x 58/73 x 57/72 x 56/71 = 0.316.

19 a) P(drug or placebo) = (120 + 100)/300 = 220/300 = 0.733

b) P(improved or not improved) = (138 + 162)/300 = 1

c) We can work this one in two ways: P(drug or improved) = (65 + 55 + 42 + 31)/300 = 193/300 = 0.643; also P(drug or improved) = P(drug) + P (improved) - P(drug and improved) = 120/300 + 138/300 - 65/300 = 193/300 = 0.643.

d) P(drug and improved) = 65/300 = 0.22.

21 a) Since the events are independent, P(A from father and A from mother) = P(A from father) x P(A from mother) = 0.75 x 0.75 = 0.5625.

b) P(Aa or aA) = P(Aa) + P(aA) = P(A) x P(a) + P(a) x P(A) = 0.75 x 0.25 + 0.25 x 0.75 = 0.375.

c) P(aa) = P(a) x P(a) = 0.25 x 0.25 = 0.0625

d)

Event	**Probability**
AA	0.5625
Aa	0.1875
aA	0.1875
aa	0.0625

e) P(Dominant trait) = P(AA) + P(Aa) + P(aA) = P(A) x P(A) + P(A) x P(a) + P(a) x P(A) = (0.75 x 0.75) + (0.75 x 0.25) + (0.25 x 0.75) = 0.9375.

23 a) p = P(HIV) = 0.03. P(at least one carries HIV) = $1 - (1 - 0.03)^6 = 1 - 0.97^6 = 0.17$.

b) p = P(HIV) = 0.03. P(at least one carries HIV) = $1 - (1 - 0.03)^{12}$

$= 1 - 0.97^{12} = 0.31$.

c) p = P(HIV) = 0.03. P(at least one carries HIV) = $1 - (1 - 0.03)^N = 1 - 0.97^N$. Since we want this expression to equal or exceed 0.50, we see that 0.97^N has to be equal to or smaller than 0.5. While this equation can be solved using logarithms, you can, with an appropriate calculator, try different values of N by keying in .97^5, .97^10, etc. until you find the lowest value of N that makes .97^N less than 0.5. That value of N is 23.

Chapter Review Exercises

1 P(Nicorette) = (43 + 109)/305 = 152/305 = 0.50

2 P(no mouth or throat soreness) = (109 + 118)/305 = 227/305 = 0.74

3 P(Nicorette or mouth/throat soreness) = P(Nicorette) + P(soreness) - P(Nicorette and soreness) = 152/305 + 78/305 - 43/305 = 187/305 = 0.61

4 P(placebo or no soreness) = P(placebo) + P(no soreness) - P(placebo and no soreness) = 153/305 + 227/305 - 118/305 = 262/305 = 0.86

5 P(soreness for the first and then soreness for the second) = P(soreness for the first) x P(soreness for the second|soreness for the first) = 78/305 x 77/304 = 0.065.

6 P(placebo for the first and the placebo for the second) = P(placebo for the first) x P(placebo for the second|Placebo for the first) = 153/305 x 152/304 = 0.25.

7
a) P(not good) = 1 - P(good) = 1 - 0.27 = 0.73.
b) Since the results of the two selections are independent of each other, P(both good) = P(first is good and second is good) = P(first is good) x P(second is good) = 0.27 x 0.27 = 0.073.
c) The expected number of good chips is 5 x 0.27 = 1.35
d) P(all five are good) = 0.27 x 0.27 x 0.27 x 0.27 x 0.27 = 0.0014; you should doubt the stated yield.

8
a) 0.10
b) No, its probability is greater than 0.05.

9
a) 1/38 = 0.026
b) Yes, the probability is less than 0.05.

10
a) P(late) = 1- P(on time) = 1 - 344/400 = 56/400 = 0.14.
b) No, the probability is greater than 0.05.

CHAPTER 7 ANSWERS

Section 7.1

1 The statement is not sensible. A correlation does not imply that one of the variables is a cause of the other.

3 The statement is sensible because a value close to +1 or -1 indicates a strong correlation.

In part (b) of Exercises 5, 7, 9, and 11, there are many reasonable answers. Please refer to Figure 7.3 of the text for guidance in drawing a scatter diagram appropriate for your answer in part (a) of each exercise.

5 Weak positive correlation since taller people tend to have larger feet.

7 Strong negative correlation since as the number of cars increases, air quality decreases.

9 Strong positive correlation since as the blood alcohol increases, reaction time also increases.

11 No correlation or weak positive correlation.

For Exercises 13, 15, 17, and 19, refer to Table 7.3 in the text.

13 Since r lies between -0.196 and -0.256, r is significant at the 0.05 significance level, but not at the 0.01 level. We conclude that there is a correlation, and there is less than a 5% chance that we would have drawn that conclusion if there really were no correlation in the population.

15 Since r lies between 0.279 and 0.361, r is significant at the 0.05 significance level, but not at the 0.01 level. We conclude that there is a correlation, and there is less than a 5% chance that we would have drawn that conclusion if there really were no correlation in the population.

17 Since r is greater than 0.561, r is significant at the 0.05 and 0.01 significance level. We conclude that there is a correlation, and there is less than a 1% chance that we would have drawn that conclusion if there really were no correlation in the population.

19 Since r is less than -0.798, r is significant at the 0.05 and 0.01 significance level. We conclude that there is a correlation, and there is less than a 1% chance that we would have drawn that conclusion if there really were no correlation in the population.

21 There is a strong positive correlation, with the correlation coefficient approximately 0.8. Much of this correlation is due to the fact that a large fraction of the grain produced is used to feed livestock.

23 a) See scatter plot at the right.

b) There is a moderate positive correlation; r = 0.59 exactly. For the actual computation of r, see the end of this chapter's solutions (p73).

c) It is difficult to argue with the phrase "no guarantee". However, four of the five lowest death rates are associated with a 55 mph speed limit. With the exception of Britain, the higher speed limits are generally associated with higher death rates. Death rates are also influenced by other factors beside speed limits.

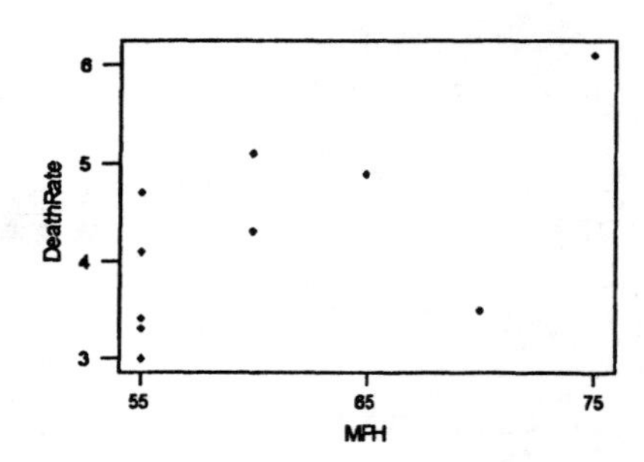

25 a) See scatter plot at the right

b) There is a moderate negative correlation (r = -0.29).

c) No.

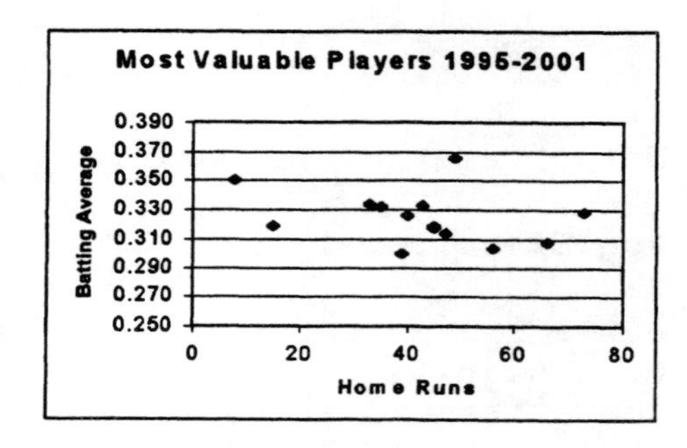

27 a) See scatter plot at the right.

b) There is a strong negative correlation between income and the number of TV hours per week (r = -0.86 exactly).

c) Families with more income have more opportunities to do other things. No.

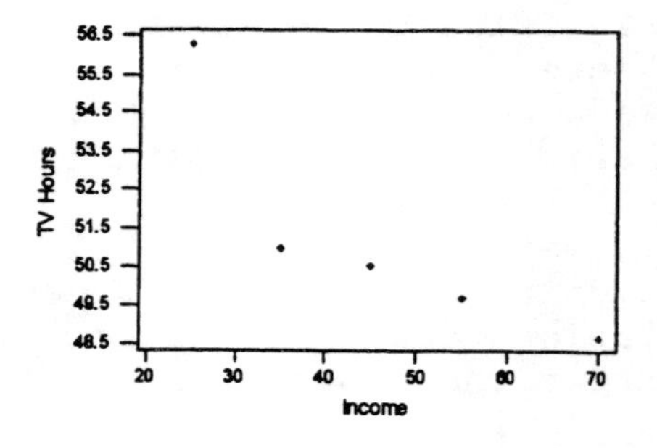

29 a) See scatter plot at the right.

b) There is a strong correlation between sales and earnings (r = 0.92 exactly). The strong correlation in this case is highly affected by the Wal-Mart data.

c) Higher sales do not necessarily translate into higher earnings. Some

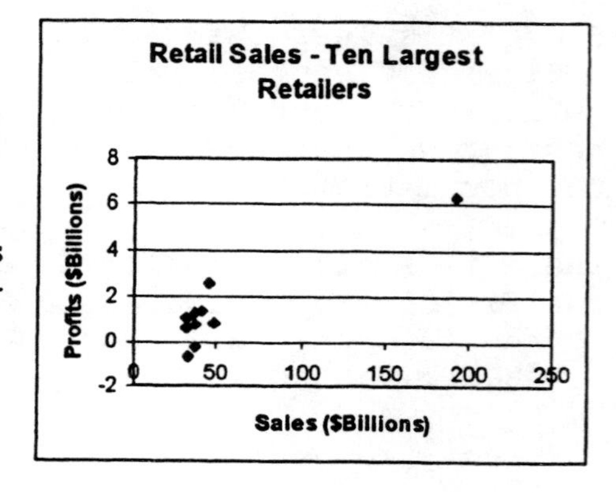

companies have larger expenses, driving earnings down.

31

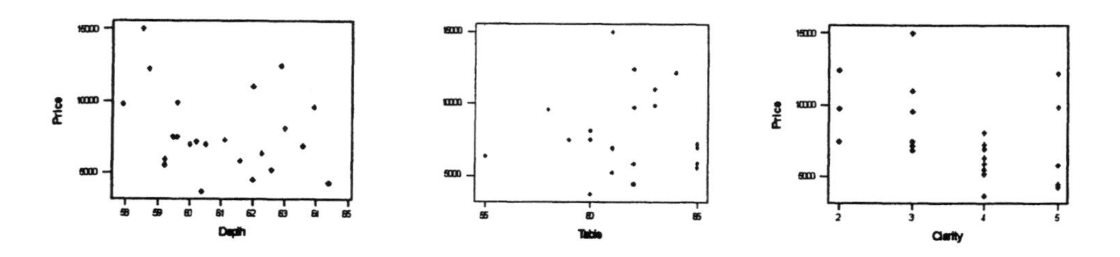

There is a moderate, but not significant, negative correlation between depth and price, no correlation between table and price, and a moderate negative correlation, significant at the 0.05 level between clarity and price. These correlations are either weaker than or opposite in direction to the correlation between weight and price. Since the correlation between clarity and price is significant, price is probably dependent on more than just weight. This helps to explain why the correlation between weight and price is not perfect.

33

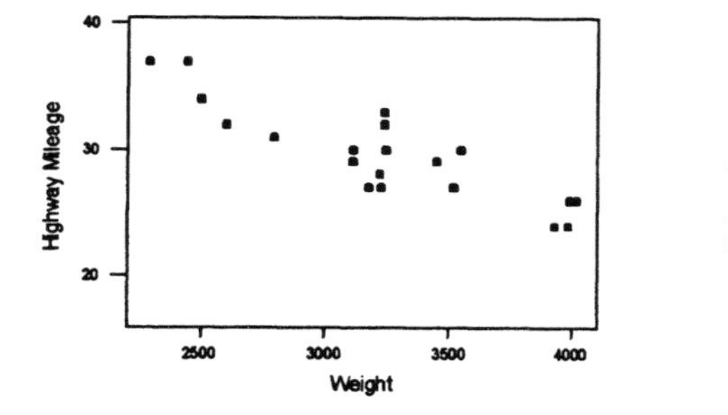

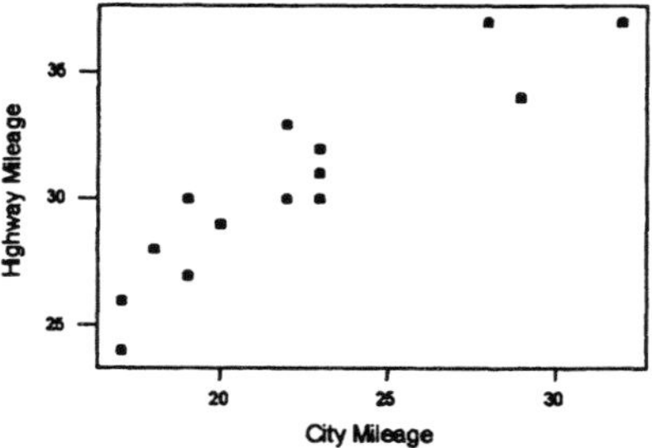

Highway mileage correlates with weight to the same extent that city mileage did ($r = -0.87$). Highway mileage and city mileage are also highly correlated ($r = 0.93$). Both correlations are significantly different from zero at the 0.01 significance level.

35 a) The variables x and y appear symmetrically in the formula (interchanging x and y does not change the formula).

b) If x (or y) is scaled (multiplied by a constant), it occurs in both the numerator and denominator; so there is no overall effect. For example, changing from pounds to ounces or kilograms or changing from miles to feet or kilometers has no effect on the correlation coefficient. If, however, the units are changed by taking logarithms, as is done in changing the concentration of hydrogen ions in an acid to pH, or by taking square roots, then the correlation coefficient *is* likely to change.

Section 7.2

1 The statement is not sensible. Although it might seem to make sense that increasing a car's weight causes the fuel consumption rate (in miles per gallon) to decrease, we cannot make that conclusion based on the correlation. The correlation does not imply a cause and effect relationship.

3 The statement is not sensible. Instead of concluding that drinking *causes* car crashes, we can only conclude that there is a relationship between drinking and car crashes. A correlation does not imply causality.

5 There is a positive correlation between the crime rate and the number of people in prison. Either a direct correlation or a common cause could explain the correlation.

7 There is a positive correlation between the miles of freeways and the congestion. A direct cause could explain the correlation or both variables could be related to the level of population in the area.

9 There is a positive correlation between grades and the ages of the students' cars. The correlation is most likely due to coincidence.

11 There is a positive correlation between gasoline prices and airline passengers. A direct cause could explain the correlation.

13 There is a positive correlation between the number ministers and priests and the movie attendance. Most likely, both are the result of a common cause, an increasing population.

15 a) The outlier is the upper left point (0.4,1.0). Without the outlier, the correlation coefficient is 0.0.
b) With the outlier, the correlation coefficient is -0.58.

17 The 1990 and 1991 data points do not appear to be consistent with the rest of the data. The correlation coefficient is slightly greater with those points removed (see the scatter plot at the right). In fact, r = 0.39 with those points included and r = 0.44 with them removed. From 1992 to 2001, there is a tendency for the unemployment rate and the inflation rate to move up and down together. Although we need to be careful about drawing conclusions without additional data, the years 1990-1992 were under the Bush presidency while 1993-2000 were under Clinton.

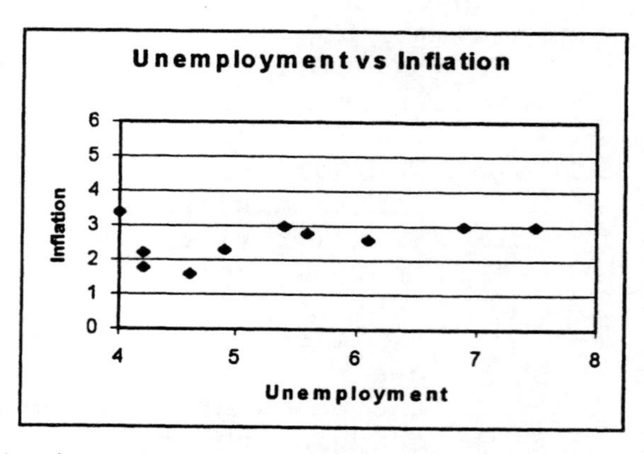

19 The actual correlation coefficient is r = 0.92, which is significant at the 0.01 level, so there is a very strong correlation between weight and shoe size.

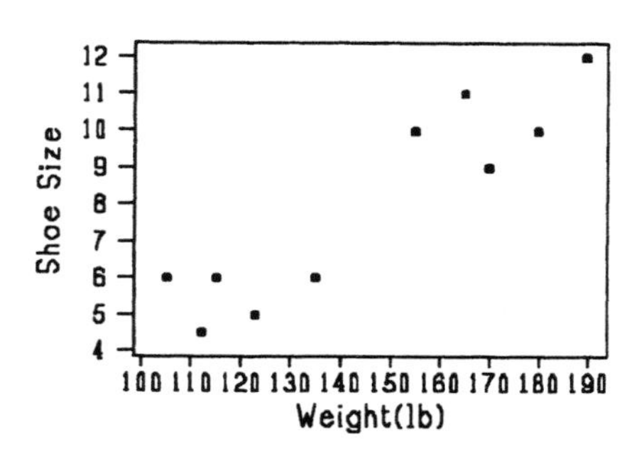

21 a) The actual correlation coefficient is r = 0.77. This is significant at the 0.01 level, indicating a strong correlation.

b) The 15 points to the left correspond to relatively poor countries, such as Uganda. The remaining points on the right correspond to relative affluent countries, such as Sweden.

c) There appears to be a negative correlation between the variables for the poorer countries and a positive correlation for the wealthier countries.

Section 7.3

1 The statement is not sensible. The projection is likely to be beyond the scope of the available data. Also, the value of a car declines with age, but some cars that are very old gain value because they become antiques.

3 The statement is not sensible. A correlation based on a sample of men does not necessarily apply to women.

5 a) See plot at the right.

b) Actual r = -0.16; r^2 = 0.026. So about 3% of the variation in price can be explained by the best-fit line.

c) The best-fit line should not be used to make predictions.

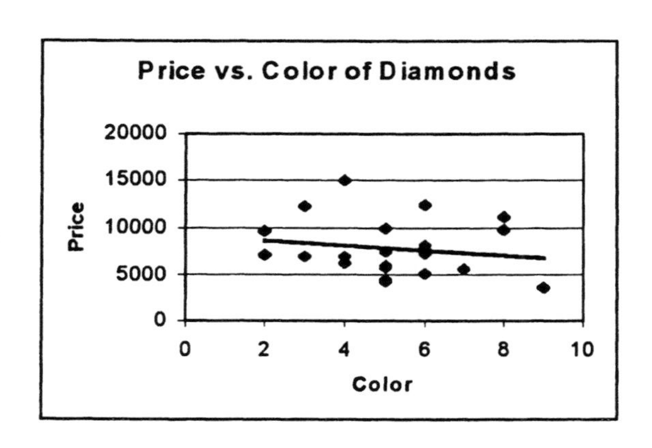

7
a) See plot at the right.
b) Actual r = -0.99; $r^2 = 0.97$. About 97% of the variation in farm size can be explained by the best-fit line.
c) The best-fit line could be used to make predictions within the range of the number of farms included in the data. Because there does appear to be a slight curvature to the points, predicting outside that range should not be done.

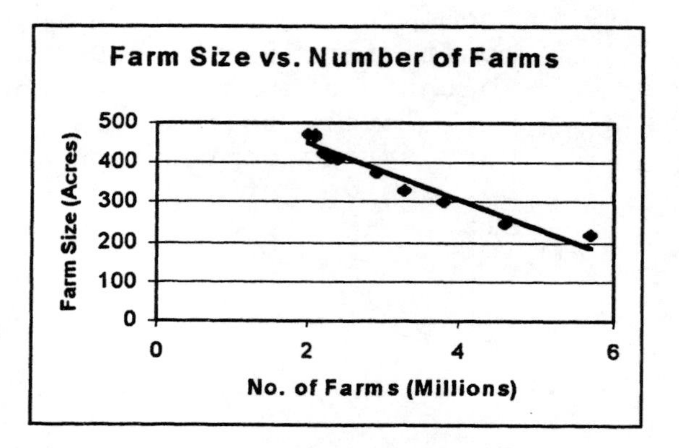

9
a) See graph at the right.
b) r = 0.052117; $r^2 = 0.0027$; about 0.27% of the variation in gross receipts is explained by the best-fit line.
c) The best-fit line should not be used to make predictions.

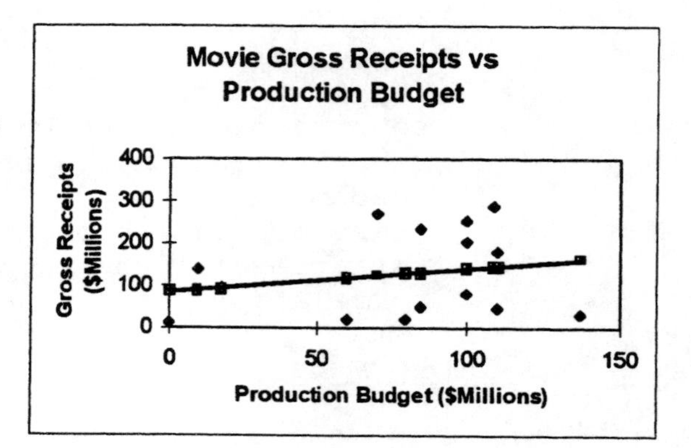

In Exercises 11, 13, 15, 17, and 19, we will show the actual best-fit lines in part (a). Your line may be different since you will have tried to draw the line by eye. Similarly, in part (b), we will show the actual value of r^2. Your estimate may be different.

11
a) See scatter plot at the right.
b) There is a moderate positive correlation; r = 0.59 exactly; $r^2 = 0.35$; 35% of the variation can be accounted for by the best-fit line. For the actual computation of r and the best-fit line, see the end of this chapter's solutions.
c) (75,6.1) and (70,3.5) are both possible outliers, the first because it is away from most of the data points and the latter because the death rate is lower than might be expected considering the rest of the data. Because one point is above the best-fit line and one is below, the net effect of the two points is probably to cancel each other

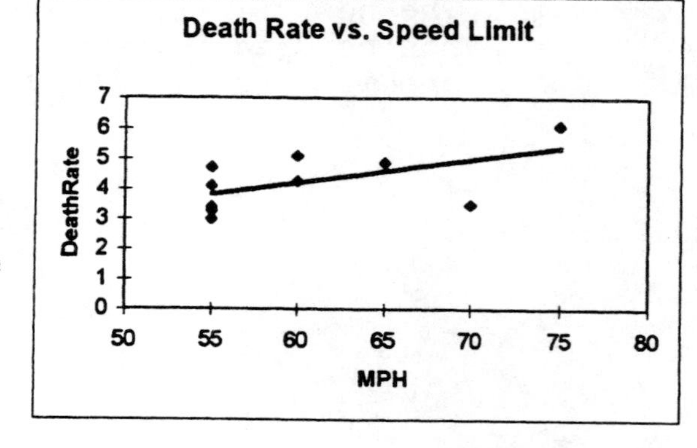

out since both points will "pull" the line toward themselves.

d) The value of r is too small to consider predictions based on the best-fit line to be reliable.

13 a) See scatter plot at the right.

b) The actual r = -0.29; r^2 = 0.08; only 8% of the variation can be accounted for by the best-fit line.

c) (.366,49) is an outlier since it is far from the other data points.

d) Predictions based on the best-fit line are not reliable.

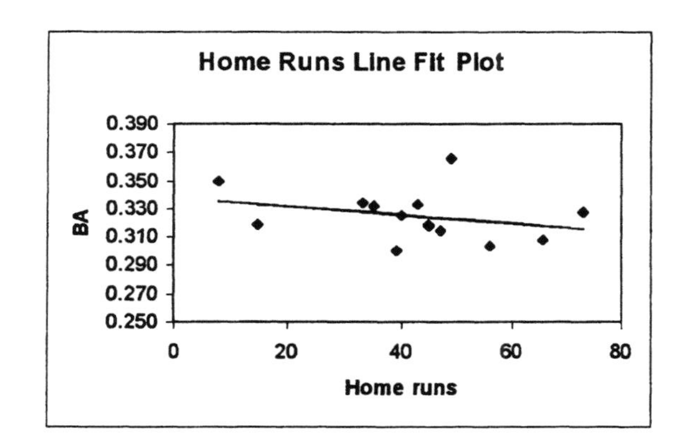

15 a) See scatter plot at the right.

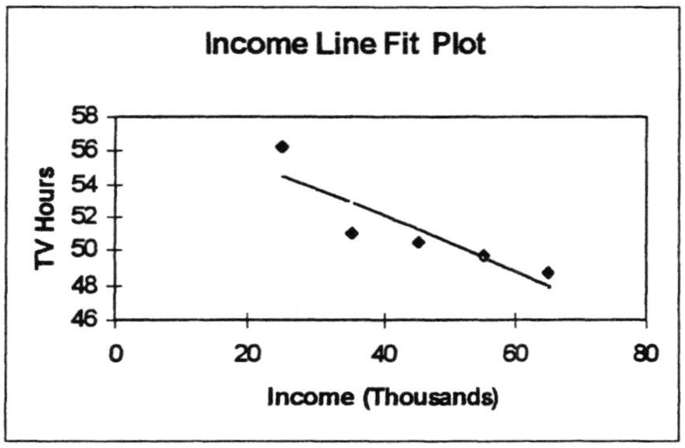

b) The actual r = -0.86; r^2 = 0.74; 74% of the variation can be accounted for by the best-fit line.

c) (25000,56.3) is an outlier. The best-fit line would have a less steep downward slope if that point were removed.

d) Predictions based on the best-fit line could be reliable, but the presence of the outlier and its effect on the best-fit line make the reliability questionable.

17 a) See scatter plot at the right.

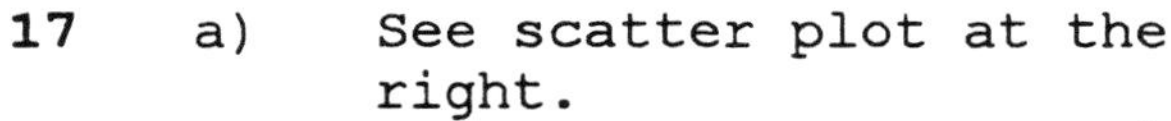

b) The actual r = 0.92; r^2 = 0.84; 84% of the variation can be accounted for by the best-fit line.

c) (193.3,6.3) is an outlier.

d) Predictions based on the best-fit line could be reliable, but the strong effect of the outlier in determining the best-fit line makes the reliability questionable. More data is needed.

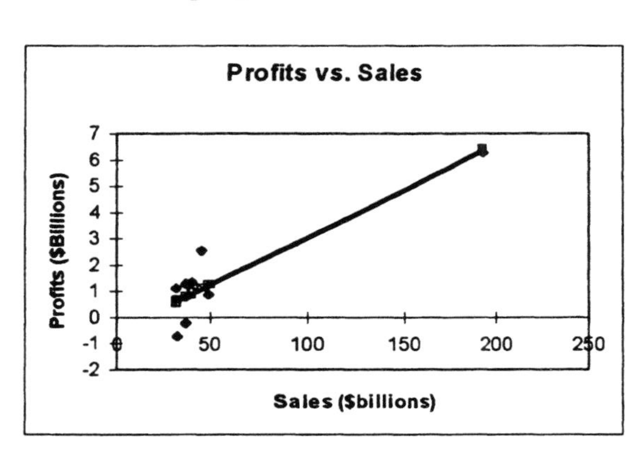

Section 7.4

1 The statement is not sensible. We cannot conclude that a correlation implies causality, regardless of the value of the correlation coefficient.

3 The statement is not sensible. Even though coincidence is ruled out, there might be a common underlying cause that could be a possible explanation.

5 The physical model involves the physics of the internal combustion engine which requires gas to run.

7 The physical model involves Newton's Law of Gravity.

9 Guideline 1
Guidelines 2 and 5
Guidelines 3 and 5
The headaches are associated with work days in some way. The headaches are not associated with Coke or possibly with caffeine. The headaches are possibly the result of bad ventilation in the building.

11 Smoking can only increase the risk already present.

13 This was an observational study. Later child bearing reflects an underlying cause. While it's possible that the conclusions are correct, there are other possible explanations for the findings. For example, it's also possible that the younger women lived during a time when having babies after age forty was less likely (by choice). It is still possible for them to live to be 100.

15 Availability is not itself a cause. Social, economic, or personal conditions cause individuals to use the available weapons.

Chapter Review Exercises

1 There appears to be a strong correlation between the interval between eruptions and the duration of the eruption.

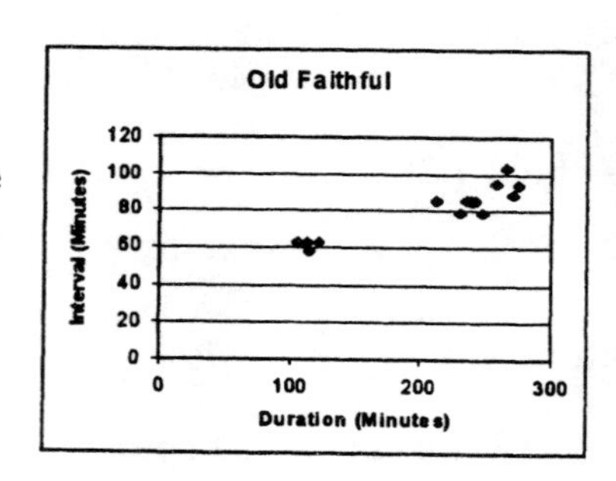

2 The height of an eruption appears to be fairly independent of the length of the interval between eruptions.

3 Since the computed value of r is greater than the table value 0.641, the correlation between interval and duration is significant at the 0.01 level. Since $r^2 = 0.87$, about 87% of the variation in interval lengths can be explained by the variation in duration.

4 Since the computer value of r is between -0.514 and +0.514, the correlation between interval and height is not significant. Since $r^2 = 0.15$, about 15% of the variation in interval lengths can be explained by the variation in height.

5 Data would consist of the death rate in a neighborhood and the distance between the neighborhood and power lines for many neighborhoods. It may be possible to establish a correlation between power lines and leukemia deaths, but it would be very difficult to establish a causal relationship.

6 The points on the scatter diagram lie on a straight line with negative slope (falling to the right).

7 Correlation alone never implies causation, and, in this case, certainly more trips to the dentist do not cause higher incomes. Households with more disposable income can afford more trips to the dentist or can afford dental insurance which covers the costs of the trips.

8 Variables affecting the value of a home might include its location, size, age, condition, and lot size. Location is often cited as the most important factor with considerations including nearness to schools, shopping, churches, etc. The age of the previous owner would be unrelated to the value.

9 The data values that were collected were uncorrelated. It's still possible that the variables represented by the data values are related in some non-linear way, i.e., the scatter plot forms a curve instead of a straight line.

Example of the Computations of r and the Best-fit Line.

We provide here the details needed for the computation of r and the best-fit line for Exercise 23 of Section 7.1 and Exercise 11 of Section 7.3. Both exercises use the same data.

Speed Limit	Death Rate			
x	y	x^2	y^2	xy
55	3.0	3025	9.00	165.0
55	3.3	3025	10.89	181.5
55	3.4	3025	11.56	187.0
70	3.5	4900	12.25	245.0
55	4.1	3025	16.81	225.5
60	4.3	3600	18.49	258.0
55	4.7	3025	22.09	258.5
65	4.9	4225	24.01	318.5
60	5.1	3600	26.01	306.0
75	6.1	5625	37.21	457.5
605	**42.4**	**37075**	**188.32**	**2602.5**

For this data, the number of data points is n = 10. The totals for each column are shown in bold at the bottom of the column.
Thus $\sum x = 605; \sum y = 42.6; \sum x^2 = 37075; \sum y^2 = 188.32; \sum (x \times y) = 2602.5$.
The formula for r is given at the end of Section 7.1.
Substituting the above values for the various sums, we have

$$r = \frac{n \times \sum(x \times y) - (\sum x) \times (\sum y)}{\sqrt{n \times (\sum x^2) - (\sum x)^2} \times \sqrt{n \times (\sum y^2) - (\sum y)^2}}$$

$$= \frac{10 \times 2602.5 - 605 \times 42.6}{\sqrt{10 \times 37075 - 605^2} \times \sqrt{10 \times 188.32 - 42.6^2}} = 0.59 .$$

$$r^2 = 0.59^2 = 0.35$$

To find the best-fit line, we use the equations found at the end of Section 7.3. Substituting the various totals from the table above into those equations, we obtain:

$$slope = m = \frac{n \times \sum(x \times y) - (\sum x) \times (\sum y)}{n \times \sum x^2 - (\sum x)^2} = \frac{10 \times 2602.5 - 605 \times 42.4}{10 \times 37075 - 605^2} = 0.07894$$

$$\text{y - intercept} = \text{b} = \frac{\sum \text{y}}{\text{n}} - m \times \frac{\sum x}{n} = \frac{42.6}{10} - 0.07894 \times \frac{605}{10} = -0.536$$

Thus the equation of the best-fit line is

$y = -0.536 + 0.07894x$ where y is the death rate and x is the speed limit.

To plot the line, we need to find two points on the line. Since any two points will suffice, we will choose $x = 55$ and $x = 75$. For $x = 55$, we have $y = -0.536 + 0.07894(55) = 3.81$, and for $x = 75$, we have $y = -0.536 + 0.07894(75) = 5.38$. The best-fit line can now be drawn by connecting the points (55,3.81) and (75,5.38).

CHAPTER 8 ANSWERS

Section 8.1

1 The statement is not sensible. A sampling error is the result of random variation, but the pollster's mistakes are in a different category.

3 The statement is sensible. The sample proportion is the best estimate of the population proportion.

5 The sample mean is $\bar{x} = \frac{12.3 + 13.1 + 14.3 + 14.2 + 13.8}{5} = \frac{67.7}{5} = 13.54$; it is a little higher than the population mean for Massachusetts.

7 The best estimate of the population mean study time is 17.8 hours. You would be more confident of this estimate if you sampled 300 students instead of 150. The sample mean becomes a more reliable estimate of the population mean as the sample size increases.

9
a) $z = \frac{\text{sample mean - population mean}}{\text{standard deviation}} = \frac{46{,}500 - 45{,}000}{1{,}100} = 1.4$, so the sample mean is 1.4 standard deviations from the population mean.
b) From Table 5.1, z = 1.4 corresponds to the 91.92 percentile, so the probability of a sample mean greater than 46,500 is 1 - .9192 = 0.0808.

11
a) The population proportion is $p = \frac{6{,}523}{12{,}345} = 0.528$.
b) The sample proportion is $\hat{p} = \frac{245}{500} = 0.49$.
c) The sample proportion is a little too low.

13 The sample proportion is $\hat{p} = \frac{73}{150} = 0.4867$. This is the best estimate of the population proportion, so we estimate that 0.4867 x 1608 = 783 people traveled from abroad. We would be more confident of the estimate if we sampled 300 people instead of 150 since the estimate becomes more reliable as the sample size increases.

15
a) $z = \frac{\text{sample proportion - population proportion}}{\text{standard deviation}} = \frac{0.32 - 0.34}{0.03} = -0.7$, so the sample proportion is 0.7 standard deviations below the mean of the sampling distribution.
b) From Table 5.1, z = -0.7 corresponds to the 24.20 percentile, so the probability of a sample proportion less than 0.32 is 0.2420.

17 Samples of size n = 1

Sample	**Mean**
A	52
C	5
G	60
M	33
O	97
Mean	**49.4**

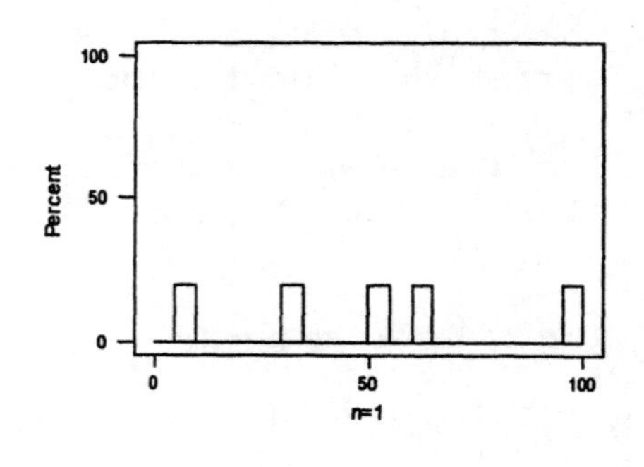

Samples of size n = 2

Sample	**Mean**
AC	28.5
AG	56.0
AM	42.5
AO	74.5
CG	32.5
CM	19.0
CO	51.0
GM	46.5
GO	78.5
MO	65.0
Mean	**49.4**

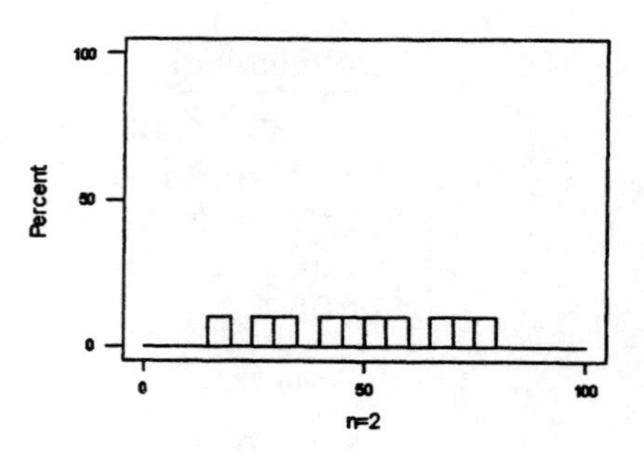

Samples of size n = 3.

Sample	**Mean**
ACG	39.00
ACM	30.00
ACO	51.33
AGM	48.33
AGO	69.67
AMO	60.67
CGM	32.67
CGO	54.00
CMO	45.00
GMO	63.33
Mean	**49.4**

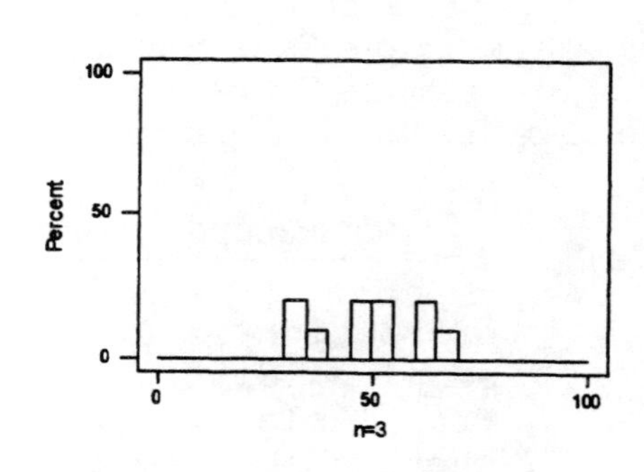

Samples of size n = 4.

Sample	Mean
ACGM	37.50
ACGO	53.50
ACMO	46.75
AGMO	60.50
CGMO	48.75
Mean	**49.4**

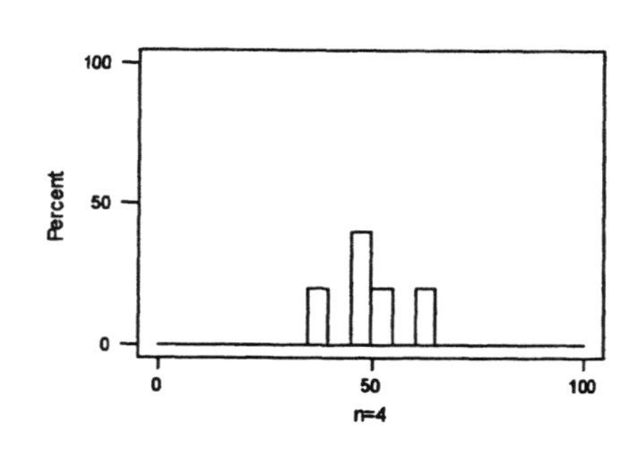

Samples of size n = 5.

Sample	Mean
ACGMO	49.4
Mean	**49.4**

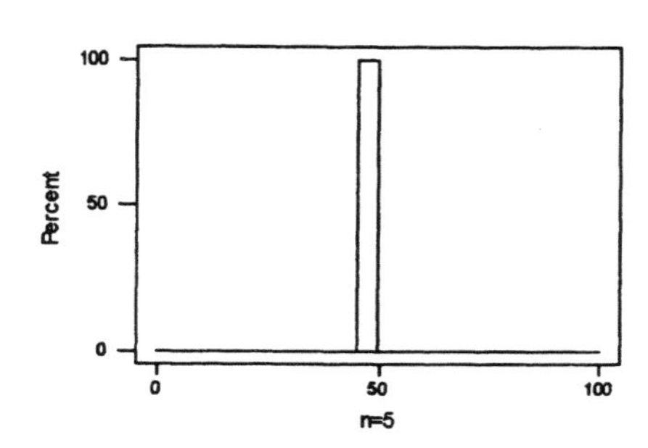

The mean of the sampling distribution of the means is 49.4 for each sample size, the same as the mean of the population.

Section 8.2

1 The statement is not sensible because the single value of 98.20 is not an interval.

3 The statement is not sensible. As the sample size increases, the margin of error tends to decrease.

5 The margin of error is $E \approx \frac{2s}{\sqrt{n}} = \frac{2 \times 10}{\sqrt{400}} = 1$. The 95% confidence interval is $\bar{x} \pm E = 120 \pm 1 = 119 \text{ to } 121$.

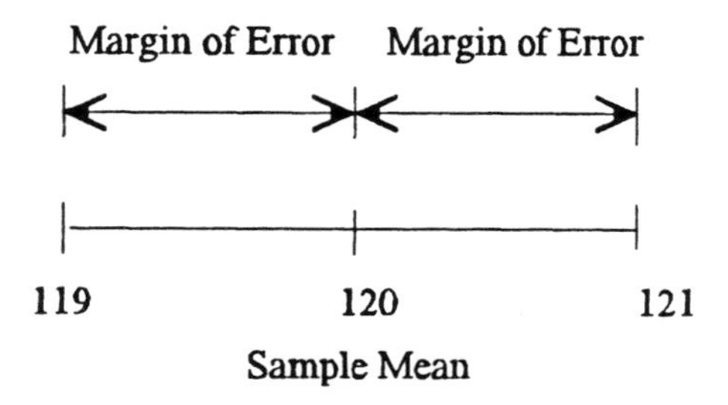

7 The margin of error is $E \approx \frac{2s}{\sqrt{n}} = \frac{2 \times 20}{\sqrt{2500}} = 0.8$. The 95% confidence interval is $\bar{x} \pm E = 160 \pm 0.8 = 159.2 \text{ to } 160.8$.

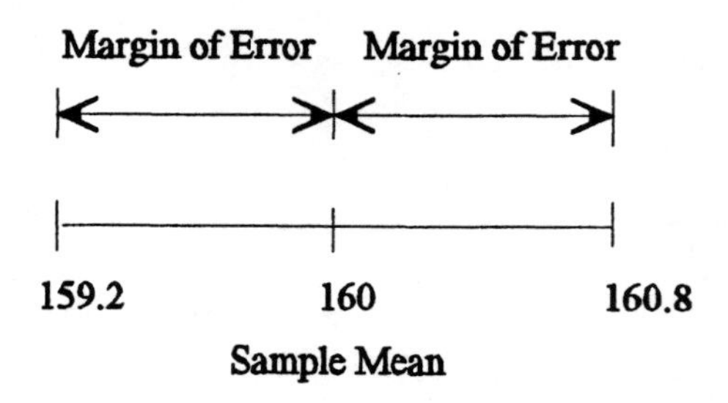

9 $n = \left(\frac{2s}{E}\right)^2 = \left(\frac{2 \times 1}{0.02}\right)^2 = 10{,}000$

11 $n = \left(\frac{2s}{E}\right)^2 = \left(\frac{2 \times 0.25}{0.01}\right)^2 = 2{,}500$

13 The margin of error is $E \approx \frac{2s}{\sqrt{n}} = \frac{2 \times 7.75}{\sqrt{81}} = 1.72$. The 95% confidence interval is $\bar{x} \pm E = \$51.25 \pm \$1.72 = \$49.53$ to $\$52.97$; $\$49.53 < \mu < \$52.97$.

15 The margin of error is $E \approx \frac{2s}{\sqrt{n}} = \frac{2 \times 1.68}{\sqrt{4400}} = .05$. The 95% confidence interval is $\bar{x} \pm E = 5.15 \pm 0.05 = 5.10$ to 5.20; $5.10 \text{ years} < \mu < 5.20 \text{ years}$.

17 For college grads, the margin of error is $E \approx \frac{2s}{\sqrt{n}} = \frac{2 \times 4500}{\sqrt{900}} = 300$. The 95% confidence interval is
$\bar{x} \pm E = 35{,}000 \pm 300 = 34{,}700$ to $35{,}300$; $\$34{,}700 < \mu < \$35{,}300$.

For high school grads, the margin of error is $E \approx \frac{2s}{\sqrt{n}} = \frac{2 \times 3900}{\sqrt{900}} = 260$. The 95% confidence interval is
$\bar{x} \pm E = 26{,}500 \pm 260 = 26{,}249$ to $26{,}760$; $\$26{,}240 < \mu < \$26{,}760$.

Neither confidence interval includes the population mean reported by the Census Bureau.

19 $n = \left(\frac{2s}{E}\right)^2 = \left(\frac{2 \times 65500}{10000}\right)^2 = 171.61 \approx 172$

21

x	$x-\bar{x}$	$(x-\bar{x})^2$	x	$x-\bar{x}$	$(x-\bar{x})^2$
2	-1.1	1.21	1	-2.1	4.41
3	-0.1	0.01	3	-0.1	0.01
6	2.9	8.41	3	-0.1	0.01
5	1.9	3.61	4	0.9	0.81
4	0.9	0.81	7	3.9	15.21
2	-1.1	1.21	3	-0.1	0.01
3	-0.1	0.01	2	-1.1	1.21
3	-0.1	0.01	3	-0.1	0.01
1	-2.1	4.41	2	-1.1	1.21
2	-1.1	1.21	2	-1.1	1.21
3	-0.1	0.01	3	-0.1	0.01
2	-1.1	1.21	4	0.9	0.81
3	-0.1	0.01	1	-2.1	4.41
4	0.9	0.81	5	1.9	3.61
5	1.9	3.61	2	-1.1	1.21
		Mean	**3.1**		**60.70**

a) The mean of the 30 data values is 93/30 = 3.1.

b) The standard deviation is $s=\sqrt{\dfrac{\sum(x-\bar{x})^2}{n-1}}=\sqrt{\dfrac{60.70}{29}}=1.447\approx 1.4$.

c) The best estimate is the sample mean, 3.1.

d) The margin of error is $E\approx\dfrac{2s}{\sqrt{n}}=\dfrac{2\times 1.4}{\sqrt{29}}=0.5$. The 95% confidence interval is $\bar{x}\pm E=3.1\pm 0.5=2.6\text{ to }3.6;\ 2.6<\mu<3.6$.

e) The sample is quite small and was concentrated in one neighborhood, so it is unlikely to be very representative of the population of all American families. It will not be very reliable.

Section 8.3

1 The statement is not sensible because the single value of 0.237 is not an interval.

3 The statement is sensible. We can see from the expression for E that larger values of n result in smaller values of E. Also, common sense suggests that larger samples are likely to result in better and more accurate (that is, smaller errors) estimates of a population proportion.

5 The margin of error is $E\approx 2\sqrt{\dfrac{\hat{p}(1-\hat{p})}{n}}=2\sqrt{\dfrac{.5(1-.5)}{400}}=0.05$. The 95% confidence interval is $\hat{p}\pm E=0.5\pm 0.05=0.45\text{ to }0.55;\ 0.45<p<0.55$.

7 The margin of error is $E \approx 2\sqrt{\frac{\hat{p}(1-\hat{p})}{n}} = 2\sqrt{\frac{.9(1-.9)}{2500}} = 0.012$. The 95% confidence interval is $\hat{p} \pm E = 0.9 \pm 0.012 = 0.888 \text{ to } 0.912;\ 0.888 < p < 0.912$.

9 $n = \frac{1}{E^2} = \frac{1}{0.02^2} = 2500$

11 $n = \frac{1}{E^2} = \frac{1}{0.04^2} = 625$

13 The margin of error is $E \approx 2\sqrt{\frac{\hat{p}(1-\hat{p})}{n}} = 2\sqrt{\frac{.35(1-.35)}{5000}} = 0.013$. The 95% confidence interval is $\hat{p} \pm E = 0.35 \pm 0.013 = 0.337 \text{ to } 0.363;\ 0.337 < p < 0.363$. In other words, the percentage lies between 33.7% and 36.3% with 95% confidence.

15 The margin of error is $E \approx 2\sqrt{\frac{\hat{p}(1-\hat{p})}{n}} = 2\sqrt{\frac{.8(1-.8)}{1400}} = 0.021$. The 95% confidence interval is $\hat{p} \pm E = 0.8 \pm 0.021 = 0.779 \text{ to } 0.821;\ 0.779 < p < 0.821$.

17 a) The margin of error is $E \approx 2\sqrt{\frac{\hat{p}(1-\hat{p})}{n}} = 2\sqrt{\frac{.172(1-.172)}{1500}} = 0.019$. The 95% confidence interval is $\hat{p} \pm E = 0.172 \pm 0.019 = 0.153 \text{ to } 0.191;\ 0.153 < p < 0.191$.

b) The margin of error is $E \approx 2\sqrt{\frac{\hat{p}(1-\hat{p})}{n}} = 2\sqrt{\frac{.132(1-.132)}{2500}} = 0.014$. The 95% confidence interval is $\hat{p} \pm E = 0.132 \pm 0.014 = 0.118 \text{ to } 0.146;\ 0.118 < p < 0.146$.

c) The margin of error is $E \approx 2\sqrt{\frac{\hat{p}(1-\hat{p})}{n}} = 2\sqrt{\frac{.229(1-.229)}{3500}} = 0.014$. The 95% confidence interval is $\hat{p} \pm E = 0.229 \pm 0.014 = 0.215 \text{ to } 0.243;\ 0.215 < p < 0.243$.

19 a) The sample proportion is 0.98.

b) From this, the margin of error is $E \approx 2\sqrt{\frac{\hat{p}(1-\hat{p})}{n}} = 2\sqrt{\frac{.98(1-.98)}{400}} = 0.014$. The 95% confidence interval is $\hat{p} \pm E = 0.98 \pm 0.014 = 0.966 \text{ to } 0.994;\ 0.966 < p < 0.994$.

c) No. To be a random sample, all films must have an equal chance of being chosen.

21 The sample proportion is 900/1600 = 0.5625. The margin of error is

$E \approx 2\sqrt{\frac{\hat{p}(1-\hat{p})}{n}} = 2\sqrt{\frac{.5625(1-.5625)}{1600}} = 0.0248$. The 95% confidence interval is 0.5625 ± .0248 = 0.538 to 0.587.

23 In the first poll, $\hat{p} = 780/1500 = 0.52$. The margin of error is $E \approx 2\sqrt{\frac{\hat{p}(1-\hat{p})}{n}} = 2\sqrt{\frac{.52(1-.52)}{1500}} = 0.0258$ and the confidence interval is 0.52 ± 0.026 = 0.494 to 0.546.
In the second poll, $\hat{p} = 1285/2500 = 0.514$. The margin of error is $E \approx 2\sqrt{\frac{\hat{p}(1-\hat{p})}{n}} = 2\sqrt{\frac{.514(1-.514)}{2500}} = 0.020$ and the confidence interval is 0.514 ± 0.020 = 0.494 to 0.534.
In the third poll, $\hat{p} = 1802/3500 = 0.515$. The margin of error is $E \approx 2\sqrt{\frac{\hat{p}(1-\hat{p})}{n}} = 2\sqrt{\frac{.515(1-.515)}{3500}} = 0.017$ and the confidence interval is 0.515 ± 0.017 = 0.498 to 0.532.
Since all of these confidence intervals include values that are less than 0.5, Martinez can not yet be confident of winning a majority of the votes.

25 a) The confidence interval is 0.54 ± 0.04 = 0.50 to 0.58 or 50% to 58%.

b) Since $E \approx 2\sqrt{\frac{\hat{p}(1-\hat{p})}{n}} = 2\sqrt{\frac{.54(1-.54)}{n}} = 0.04$, we conclude that

$$\sqrt{\frac{(.54)(.46)}{n}} = 0.02 \text{ or } \frac{(.54)(.46)}{n} = 0.0004. \text{ This implies that n} = \frac{(.54)(.46)}{0.0004} = 621.$$

27 Since $E = 2\sqrt{\hat{p}(1-\hat{p})/n}$, to decrease the margin of error by a factor of 2, we need to increase n by a factor of 4.

<u>**Chapter Review Exercises**</u>

1 a) The margin of error is $E \approx \frac{2s}{\sqrt{n}} = \frac{2 \times 3.48}{\sqrt{35}} = 1.2$. The 95% confidence interval is $\bar{x} \pm E = 134.5 \pm 1.2 = 133.3 \text{ to } 135.7;\ 133.3 < \mu < 135.7$.

b) We are 95% confident that the limits of 133.3 and 135.7 actually contain the population mean.

c) E = 1.2 millimeters from part (a).

d) The distribution of the sample means will be approximately a normal distribution.

e) Assuming that the standard deviation remained essentially unchanged, the limits would get closer together as n increases.

f) The population mean is a fixed value, not a random variable; either it is contained within the limits or it is not, and there is no associated probability. The process used to construct the confidence interval has a 95% chance of producing an interval containing the population mean. That is why we are 95% confident that the *particular* confidence interval produced contains the population mean.

2 a) $n = \left(\frac{2s}{E}\right)^2 = \left(\frac{2 \times 16}{2}\right)^2 = 256$

b) If the sample size is larger than necessary, the estimate will be better, i.e., the confidence interval will be shorter than planned. If the sample size is smaller than necessary, the estimate will be worse, i.e., the confidence interval will be longer than planned.

c) Since statistics professors are more homogeneous than the population as a whole, the standard deviation of their IQs will be less than 16. Since n is proportional to the square of s, a smaller value of s will lead to a smaller value of n.

3 a) The best estimate is the sample proportion, 111/1233 = 0.0900.

b) The margin of error is $E \approx 2\sqrt{\frac{\hat{p}(1-\hat{p})}{n}} = 2\sqrt{\frac{.09(1-.09)}{1233}} = 0.0163$. The 95% confidence interval is 0.0900 ± .0163 = 0.0737 to 0.1063.

c) We are 95% confident that the limits of 0.0737 and 0.1063 actually contains the population proportion p.

4 a) $n = \frac{1}{E^2} = \frac{1}{0.04^2} = 625$

b) No. You will end up with a sample about the right size, but it will be self-selected and likely to be biased.

CHAPTER 9 ANSWERS

Section 9.1

1 The statement is sensible. To support a claim that the technique favors girls, we must have significantly more girls than boys.

3 The statement is not sensible because the null hypothesis must include equality. The given expression can be an alternative hypothesis, but it cannot be a null hypothesis.

5 a) Not significant; this is within normal fluctuation.
b) Significant. Look for an explanation.
c) The answer varies, but perhaps 4 pounds.

7 $H_0: p = 0.11$
$H_a: p < 0.11$

9 $H_0: \mu = 78$ years
$H_a: \mu > 78$ years

11 This result is significant at the 0.05 level, indicating a bias against male babies. The P-value is 0.007 which is less than 0.05, indicating that 30 or fewer male babies in 80 births should happen only 0.7% of the time if there really is no bias.

13 This result is not significant at the 0.05 level, indicating no bias against male babies. The P-value is 0.060 which is greater than 0.05, indicating that 36 or fewer male babies in 80 births should happen 6.0% of the time if there really is no bias.

In Exercises 15 and 17, we are testing for a bias against 6s. The null hypothesis reflects the situation if there is no bias and the alternative hypothesis reflects the situation if a bias exists.

$H_0: p = 1/6$
$H_a: p < 1/6$

15 This result is significant at the 0.05 level, indicating a bias against 6s. The P-value is 0.006 which is less than 0.05, indicating that 8 or fewer 6s in 100 rolls should happen only 0.6% of the time if there really is no bias against 6s.

17 This result is not significant at the 0.05 level, indicating no bias against 6s. The P-value is 0.052 which is greater than 0.05, indicating that 12 or fewer 6s in 100 rolls should happen 5.2% of the time if there really is no bias against 6s.

Section 9.2

1 The statement is sensible because the given expression corresponds to the claim and it is a valid expression of an alternative hypothesis.

3 The statement is not sensible because the given expression corresponds

to an alternative hypothesis, not a null hypothesis.

5

$H_0: \mu =$ listed amount

$H_a: \mu >$ listed amount

Population: All packages of potato chips of the brand in question
Random sample: The answer varies. *Consumer Reports* researchers obtain samples at different times from different locations throughout the United States. Some potato chip producers proceed in the same way as part of their quality improvement/quality control process.
Conclusion: If the null hypothesis is rejected, we conclude that there is sufficient evidence to support the claim that the mean amount of preservative exceeds the amount listed on the packages.

7

$H_0: \mu =$ 1675 gal

$H_a: \mu >$ 1675 gal

Population: All households in the town
Random sample: The answer varies. One approach could be to construct a numbered list of all households, then use a computer to randomly generate numbers for households to be selected.
Conclusion: If the null hypothesis is rejected, we conclude that there is sufficient evidence to support the claim that the mean water usage among households exceeds the 1675 gallons per month.

9

$H_0: \mu(\textit{Moon}\text{ Valley}) =$ \$25,598

$H_a: \mu(\text{Moon Valley}) >$ \$25,598

Population: All adult residents of Moon Valley
Random sample: The answer varies. One approach could be to use voter registrations, motor vehicle registrations, property tax rolls, and any other lists to develop a numbered list of all adult residents, then use a computer to randomly select individuals for a sample.
Conclusion: If the null hypothesis is rejected, we conclude that there is sufficient evidence to support the claim that the mean per capita income of adult Moon Valley residents exceeds the national mean of $25,598.

11

$H_0: \mu =$ 275 lbs

$H_a: \mu >$ 275 lbs

Population: All linemen on the coach's team
Random sample: The answer varies. One approach could be to put the names of the linemen on individual index cards, put them in a bowl and mix them up, then select a sample. One could also just put their weights on the cards and proceed in the same way.
Conclusion: If the null hypothesis is rejected, we conclude that there is sufficient evidence to support the claim that the mean weight of the coach's linemen is greater than 275 pounds.

13

$H_0: \mu =$ minimum EPA level

$H_a: \mu <$ minimum EPA level

Population: All locations downstream from the plant that are not affected by other pollution sources
Random sample: The answer varies. One approach would be to partition

the downstream area into small sectors at various depths, number them, and use a computer to generate random numbers indicating the places at which samples are to be taken.
Conclusion: If the null hypothesis is rejected, we conclude that there is sufficient evidence to support the claim that the mean level of pollution is less than the critical amount specified by the EPA.

Section 9.3

1 The statement is not sensible because the test statistic and *P*-value are different measures. They will rarely be equal.

3 The statement is sensible. The *P*-value of 0.00001 indicates that the observed results are not likely to occur by chance, so we should reject the null hypothesis.

5 The standard deviation of the sampling distribution of the mean is $\frac{\sigma}{\sqrt{n}} \approx \frac{s}{\sqrt{n}} = \frac{9}{\sqrt{100}} = 0.9$; then $z = \frac{\bar{x}-\mu}{\sigma/\sqrt{n}} = \frac{24-25}{0.9} = -1.1$. The test is a left-tailed test, requiring z to be less than or equal to -1.645 in order to reject the null hypothesis and support the alternative hypothesis. Since z is not less than or equal to -1.645, we do not reject the null hypothesis at the 0.05 significance level. The alternative hypothesis is not supported.

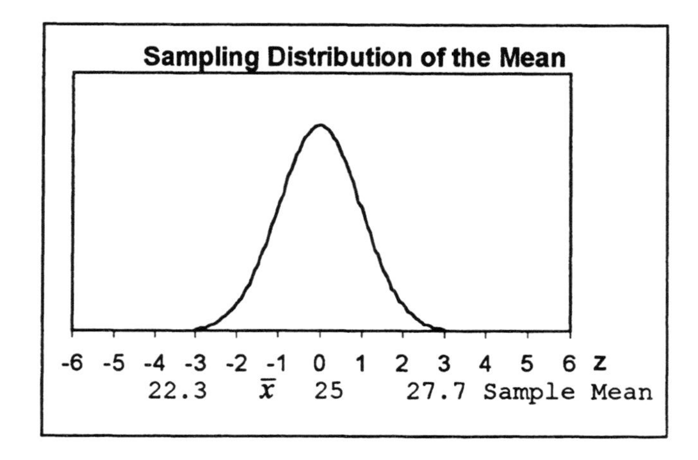

7 The standard deviation of the sampling distribution of the mean is $\frac{\sigma}{\sqrt{n}} \approx \frac{s}{\sqrt{n}} = \frac{2.3}{\sqrt{900}} = 0.0767$; then $z = \frac{\bar{x}-\mu}{\sigma/\sqrt{n}} = \frac{12.40-12.55}{0.767} = -2.0$. The test is a left-tailed test, requiring z to be less than or equal to -1.645 in order to reject the null hypothesis and support the alternative hypothesis. Since z is less than or equal to -1.645, we reject the null hypothesis at the 0.05 significance level. The alternative hypothesis is supported and we claim that the population mean is less than 12.55.

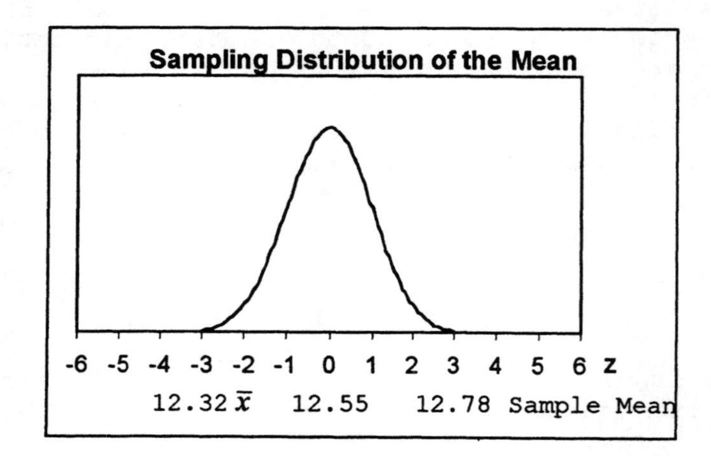

9 The standard deviation of the sampling distribution of the mean is $\frac{\sigma}{\sqrt{n}} \approx \frac{s}{\sqrt{n}} = \frac{0.1}{\sqrt{400}} = 0.005$; then $z = \frac{\bar{x} - \mu}{\sigma/\sqrt{n}} = \frac{2.045 - 2.040}{0.005} = 1.0$. The test is a right-tailed test, requiring z to be greater than or equal to 1.645 in order to reject the null hypothesis and support the alternative hypothesis. Since z is not greater than or equal to 1.645, we do not reject the null hypothesis at the 0.05 significance level. The alternative hypothesis is not supported.

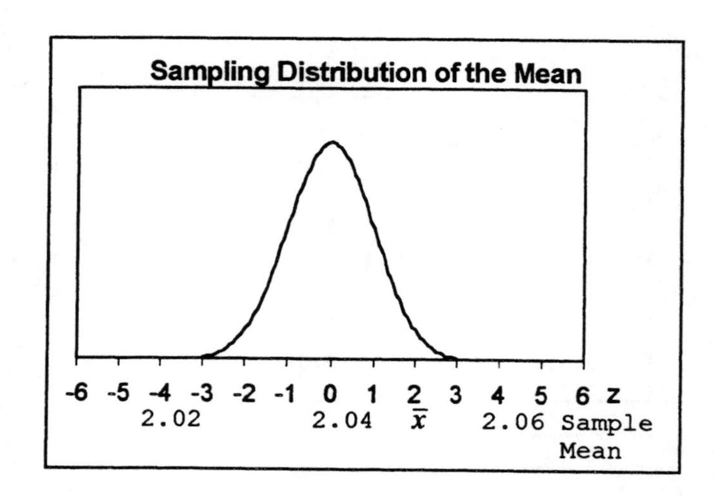

11 For z = -0.9, the P-value from Table 5.1 is 0.1841. Since this is larger than 0.05, do not reject the null hypothesis. The alternative hypothesis is not supported.

13 For z = 2.1, the P-value from Table 5.1 is 1 - 0.9821 = 0.0179. Since this is smaller than 0.05, reject the null hypothesis. The alternative hypothesis is supported.

15 For z = 2.3, the P-value from Table 5.1 is 1 - 0.9893 = 0.0107. Since this is smaller than 0.05, reject the null hypothesis. The alternative hypothesis is supported.

17 For z = -2.5, the P-value from Table 5.1 is 0.0062. Since this is smaller than 0.05, reject the null hypothesis. The alternative hypothesis is supported.

19 The sample mean is significantly *below* the assumed mean, but the claim is that the population mean is *greater* than the assumed mean. Do not reject the null hypothesis and do not support the alternative hypothesis.

21
i $H_0 : \mu = 7.5$ years; $H_a : \mu < 7.5$ years

ii The sample statistics are the sample size, n = 100, the sample mean, $\bar{x} = 7.01$ years, and the standard deviation, s = 3.74 years.

iii With s used to approximate σ, the standard deviation of the sampling distribution of the mean is $\frac{\sigma}{\sqrt{n}} \approx \frac{s}{\sqrt{n}} = \frac{3.74}{\sqrt{100}} = 0.374$ years; then

$$z = \frac{\bar{x} - \mu}{\sigma / \sqrt{n}} = \frac{7.01 - 7.50}{0.374} = -1.3 .$$

iv The mean of the sample is 1.3 standard deviations below the claimed population mean. Since -1.3 is not less than -1.645, this is not significant at the 0.05 level, and there is not sufficient evidence to support the claim that the mean time of ownership is less than 7.5 years. The P-value is 0.0968.

23
i $H_0 : \mu = 21.4$ miles per gallon; $H_a : \mu < 21.4$ miles per gallon

ii The sample statistics are the sample size, n = 40, the sample mean, $\bar{x} = 19.8$ miles per gallon, and the standard deviation, s = 3.5 mpg.

iii With s used to approximate σ, the standard deviation of the sampling distribution of the mean is $\frac{\sigma}{\sqrt{n}} \approx \frac{s}{\sqrt{n}} = \frac{3.5}{\sqrt{40}} = 0.553$ mpg; then

$$z = \frac{\bar{x} - \mu}{\sigma / \sqrt{n}} = \frac{19.8 - 21.4}{0.553} = -2.9 .$$

iv The mean of the sample is 2.9 standard deviations below the claimed population mean. Since -2.9 is less than the critical value, -1.645, this is significant at the 0.05 level, and there is sufficient evidence to support the claim that the mean for SUVs is less than 21.4 miles per gallon. The P-value is 0.0019.

25 For Alabama:

i $H_0 : \mu = \$25{,}598$; $H_a : \mu < \$25{,}598$

ii The sample statistics are the sample size, n = 900, the sample mean, $\bar{x} = \$20{,}842$, and the standard deviation, s = $12,123.

iii With s used to approximate σ, the standard deviation of the sampling distribution of the mean is $\frac{\sigma}{\sqrt{n}} \approx \frac{s}{\sqrt{n}} = \frac{12123}{\sqrt{900}} = \$404$; then

$$z=\frac{\bar{x}-\mu}{\sigma/\sqrt{n}}=\frac{\$20{,}842-\$25{,}598}{\$404}=-11.8 \quad .$$

iv The mean of the sample is 11.8 standard deviations below the claimed population mean. Since -11.8 is less than the critical value, -1.645, this is significant at the 0.05 level, and there is sufficient evidence to support the claim that the mean per capita income in Alabama is less than the national mean. The P-value is less than 0.0002.

For Illinois:

i $H_0: \mu = \$25{,}598;\ H_a: \mu > \$25{,}598$

ii The sample statistics are the sample size, n = 900, the sample mean, $\bar{x}=\$28{,}202$, and the standard deviation, s = \$18,302.

iii With s used to approximate σ, the standard deviation of the sampling distribution of the mean is $\frac{\sigma}{\sqrt{n}}\approx\frac{s}{\sqrt{n}}=\frac{18302}{\sqrt{900}}=\$610$; then

$$z=\frac{\bar{x}-\mu}{\sigma/\sqrt{n}}=\frac{\$28{,}202-\$25{,}598}{\$610}=4.3 \quad .$$

iv The mean of the sample is 4.3 standard deviations above the claimed population mean. Since 4.3 is greater than the critical value, 1.645, this is significant at the 0.05 level, and there is sufficient evidence to support the claim that the mean per capita income in Illinois is greater than the national mean. The P-value is less than 0.0002.

For Georgia:

i $H_0: \mu = \$25{,}598;\ H_a: \mu < \$25{,}598$

ii The sample statistics are the sample size, n = 900, the sample mean, $\bar{x}=\$24{,}061$, and the standard deviation, s = \$15,309.

iii With s used to approximate σ, the standard deviation of the sampling distribution of the mean is $\frac{\sigma}{\sqrt{n}}\approx\frac{s}{\sqrt{n}}=\frac{15309}{\sqrt{900}}=\$510$; then

$$z=\frac{\bar{x}-\mu}{\sigma/\sqrt{n}}=\frac{\$24{,}061-\$25{,}598}{\$610}=-3.0 \quad .$$

iv The mean of the sample is 3.0 standard deviations below the claimed population mean. Since -3.0 is less than the critical value, -1.645, this is significant at the 0.05 level; there is sufficient evidence to support the claim that the mean per capita income in Georgia is less than the national mean. The P-value is 0.0013.

For Washington:

i $H_0: \mu = \$25{,}598;\ H_a: \mu > \$25{,}598$

ii The sample statistics are the sample size, n = 900, the sample mean, $\bar{x}=\$26{,}718$, and the standard deviation, s = \$14,823.

iii With s used to approximate σ, the standard deviation of the sampling distribution of the mean is $\frac{\sigma}{\sqrt{n}} \approx \frac{s}{\sqrt{n}} = \frac{14823}{\sqrt{900}} = \$494$; then

$z = \frac{\bar{x} - \mu}{\sigma/\sqrt{n}} = \frac{\$26{,}718 - \$25{,}598}{\$494} = 2.3$.

iv The mean of the sample is 2.3 standard deviations above the claimed population mean. Since 2.3 is greater than the critical value, 1.645, this is significant at the 0.05 level, and there is sufficient evidence to support the claim that the mean per capita income in Washington is greater than the national mean. The P-value is 0.0107.

27 For April:

i $H_0: \mu = 339$DDH (degree days heating); $H_a: \mu > 339$ DDH

ii The sample statistics are the sample size, n = 50, the sample mean, $\bar{x} = 355$ DDH , and the standard deviation, s = 70 DDH.

iii With s used to approximate σ, the standard deviation of the sampling distribution of the mean is $\frac{\sigma}{\sqrt{n}} \approx \frac{s}{\sqrt{n}} = \frac{70}{\sqrt{50}} = 9.900$ DDH ; then

$z = \frac{\bar{x} - \mu}{\sigma/\sqrt{n}} = \frac{355 - 339}{9.9} = 1.6$.

iv The mean of the sample is 1.6 standard deviations above the claimed population mean. Since 1.6 is less than the critical value, 1.645, this is not significant at the 0.05 level; there is not sufficient evidence to support the claim that the mean DDH was greater than normal (339 DDH) in April of 2000. The P-value is 0.0548.

For September:

i $H_0: \mu = 69$DDH (degree days heating); $H_a: \mu < 69$ DDH

ii The sample statistics are the sample size, n = 50, the sample mean, $\bar{x} = 60$ DDH , and the standard deviation, s = 35 DDH.

iii With s used to approximate σ, the standard deviation of the sampling distribution of the mean is $\frac{\sigma}{\sqrt{n}} \approx \frac{s}{\sqrt{n}} = \frac{35}{\sqrt{50}} = 4.950$ DDH ; then

$z = \frac{\bar{x} - \mu}{\sigma/\sqrt{n}} = \frac{60 - 69}{4.95} = -1.8$.

iv The mean of the sample is 1.8 standard deviations below the claimed population mean. Since -1.8 is less than the critical value, -1.645, this is significant at the 0.05 level, and there is sufficient evidence to support the claim that the mean DDH was less than normal (69 DDH) in September of 2000. The P-value is 0.0359.

Section 9.4

1 The statement is sensible because the significance level is the probability of making a type I error.

3 The statement is not sensible. Regardless of the value of the significance level, we cannot conclude that the null hypothesis is true. We can either reject or fail to reject the null hypothesis.

5 $H_0: \mu = \text{Manufacturer's Recommended Price}$

$H_a: \mu \neq \text{Manufacturer's Recommended Price}$

Conclusion: If the null hypothesis is rejected, support the claim that the mean is not the same as the manufacturer's suggested retail price.

7 $H_0: \mu = 11{,}000 \text{ kilowatt-hours}$

$H_a: \mu \neq 11{,}000 \text{ kilowatt-hours}$

Conclusion: If the null hypothesis is rejected, support the claim that the mean is not equal to 11,000 kilowatt-hours.

9 $H_0: \mu = \text{mean blood pressure for the group with no heart problems}$

$H_a: \mu \neq \text{mean blood pressure for the group with no heart problems}$

Conclusion: If the null hypothesis is rejected, support the claim that the mean for the heart surgery patients is not the same as the mean blood pressure for the group with no heart problems.

11 a) z = 2.3 is significant. P-value = 2 x 0.0107 = 0.0214
b) z = -1.8 is not significant. P-value = 2 x 0.0359 = 0.0718
c) z = 1.2 is not significant. P-value = 2 x 0.1151 = 0.2302
d) z = -2.05 is significant. P-value = 2 x 0.0228 = 0.0456 (using z = -2.00 in Table 5.1)

13 The critical values for the test are ± 1.96. The standard deviation of the sampling distribution of the mean is $\frac{\sigma}{\sqrt{n}} \approx \frac{s}{\sqrt{n}} = \frac{9}{\sqrt{100}} = 0.9$; then

$z = \frac{\bar{x} - \mu}{\sigma/\sqrt{n}} = \frac{24.1 - 25}{0.9} = -1.0$. Since -1.0 lies between -1.96 and +1.96, do not reject the null hypothesis.

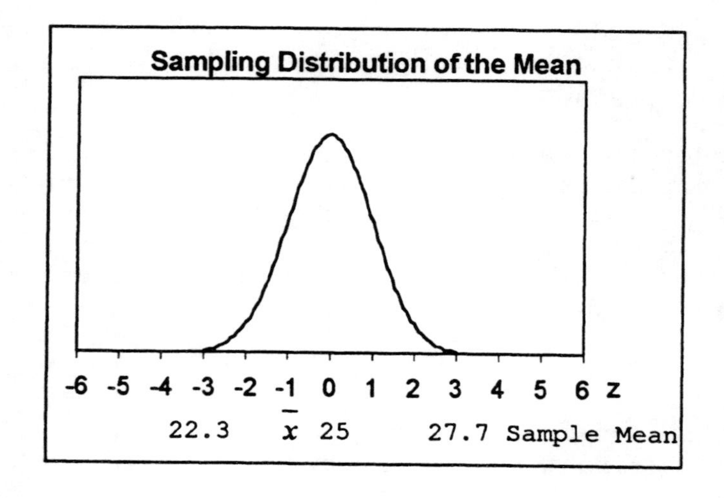

15 The critical values for the test are ± 1.96. The standard deviation of the sampling distribution of the mean is $\frac{\sigma}{\sqrt{n}} \approx \frac{s}{\sqrt{n}} = \frac{2.3}{\sqrt{900}} = 0.0767$; then

$z = \frac{\bar{x}-\mu}{\sigma/\sqrt{n}} = \frac{12.44-12.55}{0.0767} = -1.4$. Since -1.4 lies between -1.96 and +1.96, do not reject the null hypothesis.

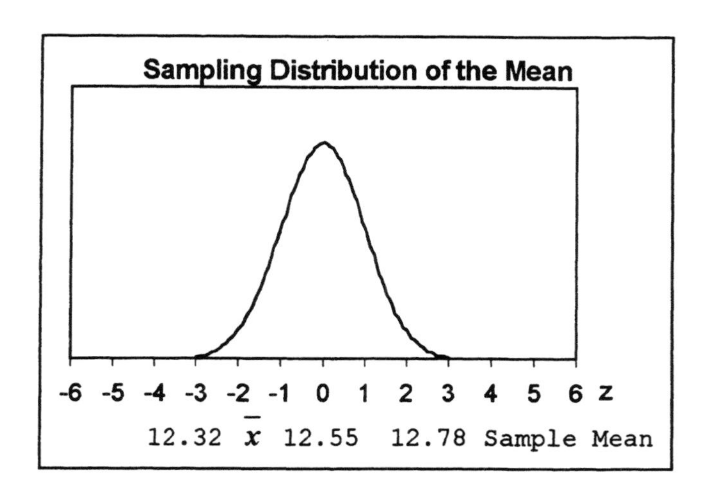

17

i $H_0: \mu = 16oz$; $H_a: \mu \neq 16oz$

ii The sample statistics are the sample size, n = 144, the sample mean, $\bar{x} = 15.8\text{oz}$, and the standard deviation, s = 1.6oz.

iii With s used to approximate σ, the standard deviation of the sampling distribution of the mean is $\frac{\sigma}{\sqrt{n}} \approx \frac{s}{\sqrt{n}} = \frac{1.6}{\sqrt{144}} = 0.133$; then

$$z = \frac{\bar{x}-\mu}{\sigma/\sqrt{n}} = \frac{15.8-16}{0.133} = -1.5 \quad .$$

iv The mean of the sample is 1.5 standard deviations below the advertised weight. Since 1.5 is between -1.96 and +1.96, this is not significant at the 0.05 level, and there is not sufficient evidence to support the claim that the mean weight of the packages is different from the advertised 16 oz. Since the probability of z being greater than 1.5 is 0.0668, the P-value is 2 x 0.0668 = 0.1336.

19

i $H_0: \mu = 600mg$; $H_a: \mu \neq 600mg$

ii The sample statistics are the sample size, n = 65, the sample mean, $\bar{x} = 589mg$, and the standard deviation, s = 21 mg.

iii With s used to approximate σ, the standard deviation of the sampling distribution of the mean is $\frac{\sigma}{\sqrt{n}} \approx \frac{s}{\sqrt{n}} = \frac{21}{\sqrt{65}} = 2.60$; then

$$z = \frac{\bar{x}-\mu}{\sigma/\sqrt{n}} = \frac{589-600}{2.60} = -4.2 \quad .$$

iv The mean of the sample is 4.2 standard deviations below the listed weight. Since -4.2 is less than -1.96, this is significant at the 0.05 level, and there is sufficient evidence to support the claim that the mean amount of acetominophen is different from the listed 2.50 cm. Since the probability of z being less than -4.2 is less than 0.0002, the P-value is less than 2 x 0.0002 = 0.0004.

21 i $H_0: \mu = 5.670g;\ H_a: \mu \neq 5.670g$

ii The sample statistics are the sample size, n = 50, the sample mean, $\bar{x} = 5.622g$, and the standard deviation, s = 0.068g.

iii With s used to approximate σ, the standard deviation of the sampling distribution of the mean is $\frac{\sigma}{\sqrt{n}} \approx \frac{s}{\sqrt{n}} = \frac{0.068}{\sqrt{50}} = 0.00962$; then

$$z = \frac{\bar{x} - \mu}{\sigma/\sqrt{n}} = \frac{5.622 - 5.670}{0.00962} = -5.0 .$$

iv The mean of the sample is 5.0 standard deviations below 5.670g. Since -5.0 is less than -1.96, this is significant at the 0.05 level, and there is sufficient evidence to support the claim that the mean weight of quarters in circulation is different from 5.67g. Since the probability of z being less than -5.0 is less than 0.0002, the P-value is less than 2 x 0.0002 = 0.0004.

23 a) Treat a patient who doesn't need treatment.
b) Fail to treat a diseased patient.
c) Treat a diseased patient.
d) Don't treat a patient who is free of disease.

25 a) Don't bet on a lottery (thinking that it is biased) when it is actually fair.
b) Bet on a lottery (thinking that it is fair) when it is actually biased.
c) Don't bet on a lottery (thinking that it is biased) when it actually is biased.
d) Bet on a lottery (thinking that it is fair) when it actually is fair.

27 a) Replace the microchip before it actually needs to be replaced
b) Leave the microchip in place even though it does need to be replaced.
c) Replace the microchip when it needs to be replaced.
d) Don't replace the microchip before it needs to be replaced.

29 a) Tamper with a production process that is actually OK.
b) Distribute tablets that do not have the correct amount of the active ingredient.
c) Correct a production process that is not working as it should be.
d) Save the time and money of not tampering with a production process that is OK.

Section 9.5

1 The statement is sensible. The null hypothesis is $p = 0.5$ and the alternative hypothesis is $p > 0.5$.

3 The statement is not sensible. In a two-tailed test, the *P*-value is either twice the area to the right of the positive *z* score or twice the area to the left of the negative *z* score.

5 a) $H_0: p = 0.915$; $H_a: p > 0.915$

b) The standard deviation of the sampling distribution of the sample proportion is computed under the assumption that the null hypothesis is true and is $\sqrt{\dfrac{p(1-p)}{n}} = \sqrt{\dfrac{0.915(1-0.915)}{100}} = 0.0279$; then $z = \dfrac{\hat{p}-p}{\sqrt{p(1-p)/n}} = \dfrac{0.920-0.915}{0.0279} = 0.2$. This value is less than the value of 1.645 required for significance at the 0.05 level; there is not sufficient evidence to conclude that there is an increase in the proportion of customers who approve of the new recipe.

c) The probability of a z value greater than 0.2 is 0.4207 and this is the P-value.

d) If the sample proportion were 93%, then the standard deviation of the distribution of the sample proportion remains the same and $z = \dfrac{\hat{p}-p}{\sqrt{p(1-p)/n}} = \dfrac{0.930-0.915}{0.0279} = 0.5$. This is still not significant.

e) If the sample proportion were 97%, then the standard deviation of the distribution of the sample proportion remains the same and $z = \dfrac{\hat{p}-p}{\sqrt{p(1-p)/n}} = \dfrac{0.970-0.915}{0.0279} = 1.97$. This is significant.

f) To be significant, the z value must equal or exceed 1.645. To find the minimum value of the sample proportion that produces significance, we first set the formula for z equal to 1.645, $z = \dfrac{\hat{p}-p}{\sqrt{p(1-p)/n}} = \dfrac{\hat{p}-0.915}{0.0279} = 1.645$. Multiplying both sides of the equation by 0.0279, and then adding 0.915 to both sides of the equation, we find that $\hat{p} = (0.0279 \times 1.645) + 0.915 = 0.961$. Since the sample size is 100, we would actually have to achieve 97 out of 100 or 0.97 for a sample proportion to be significant.

g) We follow the same procedure as in part (f), but since z must equal or exceed 2.4 for significance at the 0.01 level, we substitute 2.4 from Table 5.1 for 1.645 finding that $\hat{p} = (0.0279 \times 2.4) + 0.915 = 0.982$. Since the sample size is 100, we would actually have to achieve 99 out of 100 or 0.99 for a sample proportion to be significant. If we use a more precise table, we can use 2.326 for the minimum critical value. This leads to a minimum sample proportion of $\hat{p} = (0.0279 \times 2.326) + 0.915 = 0.9799$. Thus a sample proportion of 0.98 is actually large enough to produce significance.

7 i $H_0: p = 0.58$; $H_a: p > 0.58$

ii The sample statistics are the sample size, n = 2100, and the sample proportion 1234/2100 = 0.588.

iii The standard deviation of the sampling distribution of the sample proportion is $\sqrt{\dfrac{p(1-p)}{n}} = \sqrt{\dfrac{0.58(1-0.58)}{2100}} = 0.0108$; then

$$z = \frac{\hat{p}-p}{\sqrt{p(1-p)/n}} = \frac{0.588-0.58}{0.0108} = 0.7 .$$

iv The sample proportion is 0.7 standard deviations above p = 0.58. Since 0.7 is less than 1.645, this is not significant at the 0.05 level, and there is not sufficient evidence to support the claim that the percentage of older women at the college is above the national average. Since the probability of z being greater than 0.7 is 0.2420, the P-value is 0.2420.

9 i $H_0: p = 0.56;\ H_a: p < 0.56$

ii The sample statistics are the sample size, n = 3200, and the sample proportion 0.546.

iii The standard deviation of the sampling distribution of the sample proportion is $\sqrt{\dfrac{p(1-p)}{n}} = \sqrt{\dfrac{0.56(1-0.56)}{3200}} = 0.0088$; then

$$z = \frac{\hat{p}-p}{\sqrt{p(1-p)/n}} = \frac{0.559-0.560}{0.0088} = -0.1 .$$

iv The sample proportion is 0.1 standard deviations below p = 0.56. Since -0.1 is greater than -1.645, this is not significant at the 0.05 level, and there is not sufficient evidence to support the claim that this sample comes from a population with a married percentage less than 56%. Since the probability of z being less than -0.1 is 0.4602, the P-value is 0.4602.

11 i $H_0: p = 0.133;\ H_a: p < 0.133$

ii The sample statistics are the sample size, n = 4243, and the sample proportion 0.121.

iii The standard deviation of the sampling distribution of the sample proportion is $\sqrt{\dfrac{p(1-p)}{n}} = \sqrt{\dfrac{0.133(1-0.133)}{4243}} = 0.0052$; then

$$z = \frac{\hat{p}-p}{\sqrt{p(1-p)/n}} = \frac{0.121-0.133}{0.0052} = -2.3 .$$

iv The sample proportion is 2.3 standard deviations below p = 0.133. Since -2.3 is less than -1.645, this is significant at the 0.05 level, and there is sufficient evidence to support the claim that the poverty rate in Idaho is less than the national rate of 13.3%. Since the probability of z being less than -2.3 is 0.0107, the P-value is 0.0107.

13 i $H_0: p = 0.5;\ H_a: p > 0.5$

ii The sample statistics are the sample size, n = 1526, and the sample proportion 0.56.

iii The standard deviation of the sampling distribution of the sample proportion is $\sqrt{\dfrac{p(1-p)}{n}} = \sqrt{\dfrac{0.5(1-0.5)}{1526}} = 0.0128$; then

$$z = \frac{\hat{p} - p}{\sqrt{p(1-p)/n}} = \frac{0.56 - 0.50}{0.0128} = 4.7 \text{ .}$$

iv The sample proportion is 4.7 standard deviations above p = 0.5. Since 4.7 is greater than 1.645, this is significant at the 0.05 level, and there is sufficient evidence to support the claim that a majority of all Americans feel that gun control is an important issue. Since the probability of z being greater than 4.7 is less than 0.0002, the P-value is less than 0.0002.

15 i $H_0{:}p = 0.32$; $H_a{:}p > 0.32$

ii The sample statistics are the sample size, n = 75, and the sample proportion 0.35.

iii The standard deviation of the sampling distribution of the sample proportion is $\sqrt{\frac{p(1-p)}{n}} = \sqrt{\frac{0.32(1-0.32)}{75}} = 0.0539$; then

$$z = \frac{\hat{p} - p}{\sqrt{p(1-p)/n}} = \frac{0.35 - 0.32}{0.0539} = 0.6 \text{ .}$$

iv The sample proportion is 0.6 standard deviations above p = 0.32. Since 0.6 is less than 1.645, this is not significant at the 0.05 level, and there is not sufficient evidence to support the claim that the smoking rate for the fine arts students is higher than the national average. Since the probability of z being greater than 0.6 is 0.2743, the P-value is 0.2743.

Chapter Review Exercises

1 a) $H_0{:}\mu = 12\text{oz}$

b) $H_a{:}\mu > 12\text{oz}$

c) The standard deviation of the sampling distribution of the mean is $\frac{\sigma}{\sqrt{n}} \approx \frac{s}{\sqrt{n}} = \frac{0.11}{\sqrt{36}} = 0.0183$; then the standard score for the sample mean is $z = \frac{\bar{x} - \mu}{\sigma/\sqrt{n}} = \frac{12.19 - 12}{0.0183} = 10.4$.

d) Since the test is right-tailed, the critical value is 1.645.

e) The P-value is the probability that z is greater than 10.4, which is less than 0.0002.

f) Since the P-value is less than 0.05, we reject the null hypothesis and claim that cans of Coke have a mean amount of cola greater than 12 ounces.

g) A Type I error would result if we concluded that the population mean was greater than 12 ounces when, in fact, it was not.

h) A type II error would result if we did not conclude that the population mean was greater than 12 ounces when, in fact, it is greater than 12 ounces.

i) The P-value would be the probability that z is greater than 10.4 or less than -10.4. Since the probability that z is greater than 10.4 is less than 0.0002, the P-value is less than 2 x 0.0002 = 0.0004.

2 a) $H_0{:}p = 0.61$

b) $H_a: p \neq 0.61$

c) The sample proportion is $\hat{p} = 701/1002 = 0.700$. The standard deviation of the sampling distribution of the sample proportion is $\sqrt{\frac{p(1-p)}{n}} = \sqrt{\frac{0.61(1-0.61)}{1002}} = 0.0154$; then the standard score for the sample proportion is $z = \frac{\hat{p}-p}{\sqrt{p(1-p)/n}} = \frac{0.70-0.61}{0.0154} = 5.8$.

d) The critical values are -1.96 and +1.96.

e) The P-value would be the probability that z is greater than 5.8 or less than -5.8. Since the probability that z is greater than 5.8 is less than 0.0002, the P-value is less than 2 x 0.0002 = 0.0004.

f) Since the P-value is less than 0.05, we reject the null hypothesis and claim that the proportion who say that they voted is different from 0.61.

g) A Type I error would result if we concluded that the proportion of people who say that they voted is different from 0.61 when, in fact, it is not different from 0.61.

h) A Type II error would result if we did not conclude that the proportion of people who say that they voted is different from 0.61 when, in fact, it is different from 0.61.

i) The P-value would just be the probability that z is greater than 10.4. Since the probability that z is greater than 10.4 is less than 0.0002, the P-value is less than 0.0002.

3 The claim of an increase in the likelihood of a girl is not supported. With a sample proportion *less than* 0.5, there is no way we could ever support the claim that p > 0.5.

4 The sample is not random. It is very possible that the family is not representative of the population, so the results are biased.

CHAPTER 10 ANSWERS

Section 10.1

1 This statement is not sensible. A sensational event is not necessarily more common than a less publicized event.

3 This statement is not sensible. As you get older, your expected age of death also increases.

5 There were 41,821 fatalities over 2.7 trillion miles. Dividing 41,821 by 2,700,000,000,000, we get 0.00000001548925926 fatalities per mile. Multiplying by 100,000,000 (moving the decimal point 8 places to the right), there are about 1.5 fatalities per 100,000,000 miles.

7 There were 41,821 fatalities for a population of 185 million licensed drivers. Dividing 41,821 by 185,000,000, we get 0.0002260595 deaths per licensed driver. Multiplying by 100,000 (moving the decimal point 5 places to the right), there are about 22.6 deaths per 100,000 drivers.

9 In 1994, there were 239 fatalities. Dividing by 13.1 million flight hours, we find that there were 0.0000182 fatalities per flight hour or 1.8 fatalities per 100,000 flight hours. In 2000, there were 92 fatalities in 18.0 million flight hours. This is 0.00000511 fatalities per flight hour or 0.5 fatalities per 100,000 flight hours.

For 1994, there were 5.4 billion miles flown. Dividing this into the 239 fatalities, we find that there were 0.00000004425925926 deaths per mile or 0.044 deaths per million miles. In 2000, there were 7.1 billion miles flown. Dividing this into the 92 deaths yields 0.00000001295774648 deaths per mile or 0.013 deaths per million miles flown.

For 1994, there were 8.2 million departures. Dividing this into the 239 deaths, there were 0.00002914 deaths per departure or 2.914 deaths per 100,000 departures. For 2000, there were 11.6 million departures. The 92 deaths represent were 0.00000793 deaths per departure or 0.793 deaths per 100,000 departures.

By any of these measures, flying was less risky in 2000 than in 1994. The difference in the number of fatalities, however, could easily have been the result of one additional accident with fatalities in 1994 or the same number of accidents, but one larger plane involved in 1994.

11
a) The empirical probability of death from diabetes during a single year is 68,399/281,000,000 = 0.00024. For septicemia, it is 30,680/281,000,000 = 0.00011. The probability for death by diabetes is 2.2 times as great as for septicemia.
b) The empirical probability of death from accident during a single year is 97,860/281,000,000 = 0.000348256. For kidney disease, it is 35,525/281,000,000 = 0.000126423. The probability of death by accident is 2.75 times greater than for kidney disease.
c) The empirical probability of death from kidney disease during a single year is 35,525/281,000,000 = 0.000126423. Multiplying by 100,000, the kidney disease death rate is 12.6 per 100,000.
d) The empirical probability of death from heart disease during a single year is 725,192/281,000,000 = 0.00258075. Multiplying by 100,000, the heart disease death rate is 258 per 100,000.

e) The empirical probability of death from stroke during a single year is 167,366/281,000,000 = 0.0059560. Multiplying by 500,000, the expected number of deaths from stroke is approximately 298.

13 a) From the graph in Figure 8.12a, there are approximately 20 deaths per 1,000 people.

b) Expected number of deaths = 20/1000 x 10,200,000 = 204,000 deaths

c) From the graph in Figure 8.12b, we find the age of 40 on the horizontal axis, read up to the curve and then left to the vertical axis to the number 40. Thus a 40 year old can be expected to live another 40 years to the age of 80.

d) Following the same procedure as in part (c), we read from 60 up to the curve and then left to the number 21. Thus a 60 year old can be expected to live 11 more years to the age of 81.

e) From the graph in Figure 6.12b, a 60-year-old person has a life expectancy of 21 more years. Thus the company will receive an average of \$200 x 21 = \$4200 per person over their lifetimes. Since the death benefit is \$50,000, the company can expect to have a loss of \$45,800 per person. If it insures 1,000,000 60-year-old people, it can expect a total loss on this group of about \$45,800,000,000 over their lifetimes. We can also compute the gain or loss for just the next year. For 60-year-old people, the death rate is 20 per thousand or 0.020. Per person, the company will receive \$200 and pay out an average of \$50000 x 0.020 + \$0 x 0.980 = \$1000, leading to an average loss of \$1000 - \$200 = \$800 per person. For 1 million policy-holders, the company can expect to lose \$800 x 1,000,000 = \$800,000,000 in the next year.

15 a) Dividing 13,603 by 366 days (2000 was a leap year), Maine experienced 129 births per day, while Utah's 47,368 births were at a rate of 47,368/366 = 129 births per day.

b) Dividing 47,368 births by 1,800,000, Utah's birth rate was 0.0263 births per person or 26.3 births per 1,000 people. Dividing Maine's 13,603 births by 1,300,000, the birth rate was 0.0105 births per person or 10.5 births per 1,000 people.

17 a) There were approximately 14.8/1000 x 281,000,000 = 4,158,800 or about 4.2 million births in 2000.

b) There were approximately 8.6/1000 x 281,000,000 = 2,416,600 or about 2.4 million deaths in 2000.

c) The net population increase due to births and deaths in 2000 was about 4.2 million - 2.4 million = 1.8 million.

d) Since the total increase was actually 3.0 million, the excess increase over the number due to births and deaths is 3.0 million - 1.8 million = 1.2 million. This is the net increase due to immigration (in-migration minus out-migration). Thus the fraction of the total increase due to immigration is 1.2 million/3.0 million = 0.4, or 40%.

19 a) 7 births per 100 women (Read from the upper curve)

b) About 3.8 million live births (Read from the lower curve)

c) In 1990, there were about 4,000,000 live births and the birth rate was about 7.1 per hundred women (or 71 per thousand women). To get the number of women of child-bearing age, we reason that 4,000,000 = 71/1000 x the number of women. Thus the number of women must be 4,000,000/(71/1000) = 56,000,000. Thus there were about 56,000,000 women between the ages of 15 and 44 in 1990.

d) The peak for live births in 1960 was the baby boom, and the one in 1990 is from the children of baby boomers. The peak in the fertility rate in 1990 is a little harder to explain although it may be the result of more women in the lower half of the 15-44 age group as a result of the birth peak around 1960.

e) The fertility rate depends on the number of women ages 15-44 in the population. Thus the total births can be about the same in 1960 and 1990, while the fertility rate is lower in 1990 than in 1960 because there were more women ages 15-44 in 1990 than in 1960.

Section 10.2

1 This statement is sensible.

3 This statement is not generally true, as illustrated in the chapter by examples of confusion of the inverse.

5 Josh had the higher batting average in the first half of the season and in the second half. Overall, Josh had 85 hits in 220 at-bats, a .386 average; Jude had 80 hits in 200 at-bats, a .400 average. Thus Josh has the higher average in each half of the season, but Jude has the higher average overall, another example of Simpson's Paradox resulting because Josh had far more at-bats in the first half of the season than did Jude while Jude had far more than Josh in the second half.

7 a) New Jersey had the higher average in both racial categories. Nebraska had the higher overall overage.

b) The two states have very different racial mixes in their populations. Nebraska's overall average is very close to its average for whites since there are very few nonwhites in Nebraska's population. New Jersey's overall average is not as close to its average for whites because a much greater part of its population is nonwhite.

c) Nebraska's overall average is found by "weighting" each racial category average with the fraction of its population made up by each category. Thus the overall average is (.87)(281) + (.13)(250) = 276.97 or about 277.

d) New Jersey's overall average is found by "weighting" each racial category average with the fraction of its population made up by each category. Thus the overall average is (.66)(283) + (.34)(252) = 272.46 or about 272.

e) Simpson's Paradox arises because the two states have a greatly differing racial mix. New Jersey has the higher average score in each category, but Nebraska has the higher overall average score.

9 a) New York: The death rate for whites is 8400/4675000 = 0.001797.
The death rate for nonwhites is 500/92000 = 0.005435.
The death rate overall is 8900/74767000 = 0.0001190.

b) Richmond: The death rate for whites is 130/81000 = 0.001605.
The death rate for nonwhites is 160/47000 = 0.003404.
The death rate overall is 290/128000 = 0.002266.

c) The tuberculosis death rates for both whites and nonwhites are greater in New York than in Richmond, but the overall tuberculosis death rate is greater in Richmond than in New York. This is an example of Simpson's Paradox and it results because of the different racial mixes of the two cities, New York having a population that is about 98% white and Richmond having a population that is about 63% white.

11 a) Spellman's winning percentage at home is 10/29 = 0.345.
Morehouse's winning percentage at home is 9/28 = 0.321.
Spellman's winning percentage away is 12/16 = 0.750.
Morehouse's winning percentage away is 56/76 = 0.737.
Thus Spellman's winning percentage both at home and away is better than Morehouse's.

b) Spellman's overall winning percentage is 22/45 = 0.489.
Morehouse's overall winning percentage is 56/76 = 0.625.
Thus Morehouse College's overall winning percentage is higher than Spellman's.

c) Conference standings and championships are decided by overall records; therefore it makes more sense to claim that Morehouse College has the better team.

13 a) If 1% of the employees are drug users, there are (2000)(0.01) = 20 users. The test will correctly identify 90% of these or (20)(0.9) = 18. It will identify the other two as telling the truth.
The other 99% or 1980 of the employees do not use drugs. The test will correctly identify 90% of these or (1980)(0.9) = 1782 as telling the truth (non-users). It will incorrectly identify the remaining 10% or (1980)(0.1) = 198 as not telling the truth (users).

b) Altogether, the test identified 18 + 198 = 216 as liars. Of these, 18 were actually lying and 198 were truthful. Thus 198 out of 216 or 91.7% of those accused of lying were falsely accused.

c) Altogether, the test identified 2 + 1782 = 1784 as truthful. Of these, 1782 were actually truthful. Thus 1782 out of 1784 or 99.9% of those found truthful were actually truthful.

15 A higher percentage of women than men were hired in both the white-collar and blue-collar positions, suggesting a hiring preference for women.

Overall, 20% of the 200 females (40) who applied for white-collar positions were hired, and 85% of the 100 females (85) who applied for blue-collar positions were hired. Thus 40 + 85 = 125 of the 200 + 100 = 300 females who applied were hired, a percentage of 41.7.

Overall, 15% of the 200 males (30) who applied for white-collar positions were hired, and 75% of the 400 males (300) who applied for blue-collar positions were hired. Thus 30 + 300 = 330 of the 200 + 400 = 600 males who applied were hired, a percentage of 55.0.

Thus, overall, a higher percentage of the male applicants were hired than of the female applicants, suggesting a hiring preference for men. This is an example of Simpson's Paradox since the hiring percentage is greater for women in each job category, but overall the hiring percentage is higher for men. In this case, the positions are not alike and have different qualifications for employment. In addition, each applicant chooses the type of position for which he or she applies. It does not make sense to lump the data for the two different types of positions and therefore the data suggest either a preference for women or that a higher percentage of the women are better qualified in each position category.

17 a) In the general population, 57 + 3 = 60 of the 20,000 in the sample are infected. This is an incidence rate of 60/20000 = 0.003 or 0.3%. In the "at-risk" population, 475 + 25 = 500 of the 5,000 in the sample are infected. This is an incidence rate of 500/5000 = 10.0%.

b) In the "at-risk" category, 475 out 500 infected with HIV test positive, or 95%. Of those who test positive, 475 out of 475 + 225 = 700 have HIV, a percentage of 475/700 = 0.679 or 67.9%. These two figures are different because they measure different things. The 700 who test positive include 225 who were false positives. While the test correctly identifies 95% of those who have HIV, it also incorrectly identifies some who do not have HIV. Thus only 67.9 % of those who test positive actually have HIV.

c) In the "at-risk" population, a patient who tests positive for the disease has about a 68% chance of actually having the disease. This is nearly 7 times as great as the incidence rate (10%) of the disease in the at-risk category. Thus the test is very valuable in identifying those with HIV.

d) In the general population, patients with HIV test positive 57 times out of 60, or 95% of the time. Of those who test positive, 57 out of 57 + 997 = 1054 actually have HIV, a percentage of 57/1054 = 0.054 or 5.4%. These two figures are different because they measure different things. The 1054 who test positive include 997 who were false positives. While the test correctly identifies 95% of those who have HIV, it also incorrectly identifies some who do not have HIV. Thus only 5.4 % of those who test positive actually have HIV.

e) In the general population, a patient who tests positive for the disease has about a 5.4% chance of actually having the disease. This is 18 times as great as the incidence rate (0.3%) of the disease in the general population. Thus the test is very valuable in identifying those with HIV.

Section 10.3

1 This statement is not sensible. A two-way table is used to describe a collection of people or objects according to two variables.

3 This conclusion is not sensible. If gender and citizenship were unrelated, we would expect 5 people in each category (male/female, citizen/non-citizen); zero female citizens suggests that the categories are not independent.

5 This statement is not sensible. A large χ^2 implies that there are large differences between observed and expected values, which means the variables are not independent.

7 Column A must sum to 65. Therefore cell AD = 18.
Row C must sum to 56. Therefore cell BC = 9. The remaining Row D total, Column C total, and grand total can now be completed to yield the table below:

	A	B	Total
C	47	9	56
D	18	25	41
Total	65	32	97

9

a) The column totals are 85 and 63; the row totals are 84 and 64. Both the row totals and the column totals sum to 148, the grand total.

b) Women make up 84/148 = 0.568 or 56.8% of the sample.

c) Part-time students make up 64/148 = 0.426 or 42.6% of the sample.

d) P(man) = 64/148 = 0.432.

e) Part-timers make up 37/84 = 0.440 or 44.0% of the women.

f) P(a man is full-time) = 38/64 = 0.587

g) Women make up 378/63 = 0.587 or 58.7% of part-time students.

h) P(a full-time student is a man) = 38/85 = 0.447.

i) No. For example, 37/84 = 0.440 of women are part-time while 26/64 = 0.406 of men are part-time.

11

a) The column totals are 343 and 11; the row totals are 156 and 198. Both the row totals and the column totals sum to 354, the grand total.

b) Oregonians make up 156/354 = 0.441 or 44.1% of the sample.

c) Registered motorcycles make up 11/354 = 0.031 or 3.1% of the sample.

d) P(Connecticut resident registered an automobile) = 193/198 = 0.975.

e) Registered motorcycles make up 6/156 = 0.038 or 3.8% of the Oregon registrations.

f) P(motorcycle is registered in Connecticut) = 5/11 = 0.455.

g) Oregon automobiles make up 150/343 = 0.437 or 43.7% of the registered automobiles.

h) The overlapping data would allow some people to be counted twice.

13

a) The column totals are 15 and 24; the row totals are 20 and 19. Both the row totals and the column totals sum to 39, the grand total.

b) P(man) = 20/39 = 0.513; P(woman) = 19/39 = 0.487;

c) P(soccer game) = 24/39 = 0.615; P(play) = 15/39 = 0.385

d) P(man and play) = P(man) x P(play) = 0.513 x 0.385 = 0.197

e) For man and play, expected frequency = $39 \times \frac{20}{39} \times \frac{15}{39} = 7.692$

For man and soccer, expected frequency = $39 \times \frac{20}{39} \times \frac{24}{39} = 12.308$

For woman and play, expected frequency = $39 \times \frac{19}{39} \times \frac{15}{39} = 7.308$

For woman and soccer, expected frequency = $39 \times \frac{19}{39} \times \frac{24}{39} = 11.692$

f) The differences are not very great, all being 1.692 in absolute value.

15

	Improvement	No Improvement	Total
Drug	56	42	98
Placebo	49	43	92
Total	105	85	190

For Drug and Improvement, expected frequency = $190 \times \frac{98}{190} \times \frac{105}{190} = 54.158$

For Drug and No Improvement, expected frequency = $190 \times \frac{98}{190} \times \frac{85}{190} = 43.842$

For Placebo and Improvement, expected frequency = $190 \times \frac{92}{190} \times \frac{105}{190} = 50.842$

For Placebo and No Improvement, expected frequency = $190 \times \frac{92}{190} \times \frac{85}{190} = 41.158$

These are summarized in the following table of expected frequencies:

	Improvement	No Improvement	Total
Drug	54.158	43.842	98
Placebo	50.842	41.158	92
Total	105	85	190

The differences between observed and expected frequencies are very small.

17

	Voter	No vote	Total
Women	139	111	250
Men	132	118	250
Total	271	229	500

For Women Voters, expected frequency = $500 \times \frac{271}{500} \times \frac{250}{500} = 135.5$

For Women Nonvoters, expected frequency = $500 \times \frac{229}{500} \times \frac{250}{500} = 114.5$

For Men Full-time, expected frequency = $500 \times \frac{271}{500} \times \frac{250}{500} = 135.5$

For Men Part-time, expected frequency = $500 \times \frac{229}{500} \times \frac{250}{500} = 114.5$

These are summarized in the following table of expected frequencies:

	Voter	No vote	Total
Women	135.5	114.5	250
Men	114.5	135.5	250
Total	271	229	500

The differences between observed and expected frequencies are small.

19

	White	African American	Native American	Asian/Pacific Islands	Hispanic Origin	Total
Nebraska	1555	66	15	21	68	1725
Nevada	1448	125	30	74	253	1930
Total	3003	191	45	95	321	3655

For Nebraska Whites, expected frequency = $3655 \times \frac{1725}{3655} \times \frac{3003}{3655} = 1417.285$

For Nebraska Af. Am., expected frequency = $3655 \times \frac{1725}{3655} \times \frac{191}{3655} = 90.143$

For Nebraska Nat. Am., expected frequency = $3655 \times \frac{1725}{3655} \times \frac{45}{3655} = 21.238$

For Nebraska As. Pac Is., expected frequency = $3655 \times \frac{1725}{3655} \times \frac{95}{3655} = 44.836$

For Nebraska Hisp. Orig., expected frequency = $3655 \times \frac{1725}{3655} \times \frac{321}{3655} = 151.498$

For Nevada Whites, expected frequency = $3655 \times \frac{1930}{3655} \times \frac{3003}{3655} = 1585.715$

For Nevada Af. Am., expected frequency = $3655 \times \frac{1930}{3655} \times \frac{191}{3655} = 100.856$

For Nevada Nat. Am., expected frequency = $3655 \times \frac{1930}{3655} \times \frac{45}{3655} = 23.762$

For Nevada As. Pac. Is., expected frequency = $3655 \times \frac{1930}{3655} \times \frac{95}{3655} = 50.164$

For Nevada Hisp. Orig., expected frequency = $3655 \times \frac{1930}{3655} \times \frac{321}{3655} = 169.502$

These are summarized in the following table of expected frequencies:

	White	African American	Native American	Asian/Pacific Islands	Hispanic Origin	Total
Nebraska	1417.285	90.143	21.238	44.836	151.498	1725
Nevada	1585.715	100.856	23.762	50.164	169.502	1930
Total	3003.000	191.000	45.000	95.000	321.000	3655

The differences between observed and expected frequencies appear to be quite large.

21

	20-24	25-34	35-44	45-64	Total
Smokers	18	15	17	15	65
Nonsmokers	32	35	33	35	135
Total	50	50	50	50	200

For smokers in each age category, expected frequency =

$200 \times \frac{65}{200} \times \frac{50}{200} = 16.25$

For nonsmokers in each age category, expected frequency =

$200 \times \frac{135}{200} \times \frac{50}{200} = 33.75$

These are summarized in the following table of expected frequencies:

	20-24	25-34	35-44	45-64	Total
Smokers	16.25	16.25	16.25	16.25	65
Nonsmokers	33.75	33.75	33.75	33.75	135
Total	50	50	50	50	200

The differences between observed and expected frequencies appear to be small.

23 a) H_0: Getting a flu shot is independent of getting the flu.
H_a: Getting a flu shot and getting the flu are dependent.

b) For shot and flu, the expected frequency $= 500 \times \frac{300}{500} \times \frac{200}{500} = 120$.

For shot and no flu, the expected frequency = $500 \times \frac{300}{500} \times \frac{300}{500} = 180$.

For no shot and flu, the expected frequency $= 500 \times \frac{200}{500} \times \frac{200}{500} = 80$.

For no shot and no flu, the expected frequency = $500 \times \frac{200}{500} \times \frac{300}{500} = 120$

The data and these results are summarized in the following table of observed and expected frequencies:

	Flu	No Flu	Total
Flu Shots	100(120)	200(180)	300
No Flu Shots	100(80)	100(120)	200
Total	200	300	500

c) The calculation of χ^2 is shown in the following table:

Outcome	O	E	O-E	$(O-E)^2$	$(O-E)^2/E$
Shot, Flu	100	120	-20	400	3.333
Shot, No Flu	200	180	20	400	2.222
No Shot, Flu	100	80	-20	400	5.000
No Shot, no Flu	100	120	-20	400	3.333
Totals	500	500	0		$\chi^2 = 13.889$

d) From Table 10.12, the critical value for a 2 x 2 table is 6.635 for the 0.01 significance level. Since χ^2=13.889 exceeds this value, we reject the null hypothesis at the 0.01 level and conclude that getting a flu shot and getting the flu are dependent.

25 a) H_0: Age category and drinking status are independent.
H_a: Age category and drinking status are dependent.

b) The row and column totals are shown in the table below.
For Age < 25 and drinking, the expected frequency

$$= 3281 \times \frac{1252}{3281} \times \frac{50}{3281} = 19.080 .$$

For Age < 25 and non-drinking, the expected frequency =

$$3281 \times \frac{1252}{3281} \times \frac{3231}{3281} = 1232.920$$

For Age $\geq$ 25 and drinking, the expected frequency =

$$3281 \times \frac{2029}{3281} \times \frac{50}{3281} = 30.920 .$$

For Age $\geq$ 25 and non-drinking, the expected frequency =

$$3281 \times \frac{2029}{3281} \times \frac{3231}{3281} = 1998.080 .$$

The data and these results are summarized in the following table of observed and expected frequencies:

	Drinking	Non-drinking	Total
Age < 25	13(19.080)	1239(1232.920)	1252
Age $\geq$ 25	37(30.920)	1992(1998.080)	2029
Total	50	3231	3281

c) The calculation of χ^2 is shown in the following table:

Outcome	O	E	O-E	$(O-E)^2$	$(O-E)^2/E$
Age > 25, Drinking	13	19.08	-6.08	36.9664	1.9374
Age > 25, Non-Drinking	1239	1232.92	6.08	36.9664	0.0300
Age ≥ 25, Drinking	37	30.92	6.08	36.9664	1.1955
Age ≥ 25, Non-Drinking	1992	1998.08	-6.08	36.9664	0.0185
Totals	3281	3281			$\chi^2 = 3.1814$

d) From Table 10.12, the critical value for a 2 x 2 table is 3.841 for the 0.05 significance level. Since χ^2=3.1814 does not exceed this value, do not reject independence at the 0.05 level. It does not appear that age category and drinking status are dependent.

27 a) H_0: Poverty status is independent of whether a person has a high school diploma.
H_a: Poverty status and whether a person has a high school diploma are dependent.

b) For Poverty and high school diploma, the expected frequency $= 345 \times \frac{38}{345} \times \frac{69}{345} = 7.6$.

For Poverty and no high school diploma, the expected frequency = $345 \times \frac{38}{345} \times \frac{276}{345} = 30.4$.

For No Poverty and high school diploma, the expected frequency = $345 \times \frac{307}{345} \times \frac{69}{345} = 61.4$.

For No Poverty and no high school diploma, the expected frequency $= 345 \times \frac{307}{345} \times \frac{276}{345} = 245.6$.

The data and these results are summarized in the following table of observed and expected frequencies:

	H.S. Diploma	No H.S. Diploma	Total
Poverty	23(7.6)	15(30.4)	38
No Poverty	46(61.4)	261(245.6)	307
Total	69	276	345

c) The calculation of χ^2 is shown in the following table:

Outcome	O	E	O-E	(O-E)²	(O-E)²/E
Poverty, H.S. Diploma	23	7.6	15.4	237.16	31.205
Poverty, No H.S. Diploma	15	30.4	-15.4	237.16	7.801
No Poverty, H.S. Diploma	46	61.4	-15.4	237.16	3.863
No Poverty, No H.S. Diploma	261	245.6	15.4	237.16	0.966
Totals	345	345			$\chi^2 = 43.835$

d) From Table 10.12, the critical value for a 2 x 2 table is 6.635 for the 0.01 significance level. Since $\chi^2 = 43.835$ exceeds this value, reject independence at the 0.01 level. It appears that poverty status and whether a person gets a high school diploma are dependent.

29 a) H_0: The sport is independent of whether the wins are at home or away.
H_a: The sport and whether the wins are at home or away are dependent.

b) For basketball and home team wins, the expected frequency =
$$490 \times \frac{198}{490} \times \frac{287}{490} = 115.971 .$$
For basketball and visiting team wins, the expected frequency =
$$490 \times \frac{198}{490} \times \frac{203}{490} = 82.029 .$$

For baseball and home team wins, the expected frequency =
$$490 \times \frac{100}{490} \times \frac{287}{490} = 58.571 .$$
For baseball and visiting team wins, the expected frequency =
$$490 \times \frac{100}{490} \times \frac{203}{490} = 41.429 .$$
For hockey and home team wins, the expected frequency =
$$490 \times \frac{93}{490} \times \frac{287}{490} = 54.471 .$$
For hockey and visiting team wins, the expected frequency =
$$490 \times \frac{93}{490} \times \frac{203}{490} = 38.529 .$$
For football and home team wins, the expected frequency =
$$490 \times \frac{99}{490} \times \frac{287}{490} = 57.986 .$$

For football and visiting team wins, the expected frequency =
$490 \times \frac{99}{490} \times \frac{203}{490} = 41.014$.

The data and these results are summarized in the following table of observed and expected frequencies:

	Home Team Wins	Visiting Team Wins	Total
Basketball	127(115.971)	71(82.029)	198
Baseball	53(58.571)	47(41.429)	100
Hockey	50(54.471)	43(38.529)	93
Football	57(57.986)	42(41.014)	99
Total	287	203	490

c) The calculation of χ^2 is shown in the following table:

Outcome	O	E	O-E	$(O-E)^2$	$(O-E)^2/E$
Basketball, Home Wins	127	115.971	11.021	121.462	1.049
Basketball, Visitor Wins	71	82.029	-11.021	121.462	1.483
Baseball, Home Wins	53	58.571	-5.571	31.036	0.530
Baseball, Visitor Wins	47	41.429	5.571	31.036	0.749
Hockey, Home Wins	50	54.471	-4.471	19.990	0.367
Hockey, Visitor Wins	43	38.529	4.471	19.990	0.519
Football, Home Wins	57	57.986	-0.986	0.972	0.017
Football, Visitor Wins	42	41.014	0.986	0.972	0.024
Totals	490	490			$\chi^2 = 4.737$

d) From Table 10.12, the critical value for a 4 x 2 table is 7.815 for the 0.05 significance level. Since χ^2= 4.737 does not exceed this value, do not reject independence at the 0.05 level. It does not appear that the sport and home/away wins are related.

Chapter Review Exercises

1 a) Assuming a U.S. population of 281 million (See Exercise 17 in Section 10.1), the overall death rate due to accidents is 97,000/281,000,000 = 0.0003451957295 per person. Multiplying by 100,000 (moving the decimal point 5 places to the right), the death rate per 100,000 people is 34.52.

b) Of the 97,000 accidental deaths, 25,000 involved people 75 years of age or older; therefore 72,000 accidental deaths involved people younger than 75. To get the probability of a person dying by an accident and being less than 75 years old, we divide 72,000 by the size of the population, 281,000,000, yielding a probability of 0.000256.

c) The ratio 42000/17000 = 2.47; thus you are approximately 2.5 times as likely to die in an automobile accident than in a fall.

d) Again assuming a U.S. population of 281,000,000, the death rate for falls is 17,000/281,000,000 = 0.00006049822064 per person. Multiplying by 1,000,000 (moving the decimal point 6 places to the right, the fall death rate per million people is about 60.50.

2 a) Among the 100,000 people who pass through the scanning device each week, 1% (or 1000) carry such metal objects. The other 99,000 do not carry such objects. Ninety-eight percent of these 99,000 non-carrying people (or 97,020) are correctly identified as non-carriers. The other 2%, 1980 non-carrying people, are identified incorrectly as carriers (false positives).

b) From part a, 97,020 non-carriers are identified correctly as non-carriers (true negatives).

c) Of the 1000 carriers, 98% or 980 are identified as carriers. The other 20 are identified incorrectly as non-carriers (false negatives).

d) From part c, 980 carriers are identified correctly as carriers (true positives).

e) Altogether 980 carriers are identified as carriers and 1980 non-carriers are identified as carriers. Thus a total of 2960 people are identified as carriers. Thus 1980/2960 = 0.6689 or 66.89% of those identified as carriers are falsely identified.

3 Creating the table below will be helpful.

	Regular Priced Items	**Advertised-Special Items**	**Total**
Undercharge	20	7	27
Overcharge	15	29	44
Correct Price	384	364	748
Total	419	400	819

a) Among regular priced items, 15/384 = 0.0358 or 3.58% are overcharged.

b) H_0: Charge and pricing are independent.
H_a: Charge and pricing are dependent.

c) For regular price and undercharge, the expected frequency =

$$819\times\frac{27}{819}\times\frac{419}{819}=13.813$$

For advertised specials and undercharge, the expected frequency =

$$819\times\frac{27}{819}\times\frac{400}{819}=13.187$$

For regular price and overcharge, the expected frequency =

$$819 \times \frac{44}{819} \times \frac{419}{819} = 22.510$$

For advertised specials and overcharge, the expected frequency =

$$819 \times \frac{44}{819} \times \frac{400}{819} = 21.490$$

For regular price and correct charge, the expected frequency =

$$819 \times \frac{748}{819} \times \frac{419}{819} = 382.676$$

For advertised specials and correct charge, the expected frequency

$$= 819 \times \frac{748}{819} \times \frac{400}{819} = 365.324$$

The data and the expected frequencies are summarized in the following table:

	Regular Priced Items	Advertised-Special Items	Total
Undercharge	20(13.813)	7(13.187)	27
Overcharge	15(22.510)	29(21.490)	44
Correct Price	384(382.676)	364(365.324)	748
Total	419	400	819

d) The calculation of χ^2 is shown in the following table.

Outcome	O	E	O-E	$(O-E)^2$	$(O-E)^2/E$
Undercharge, Regular	20	13.813	6.187	38.279	2.771
Undercharge, Special	7	13.187	-6.187	38.279	2.903
Overcharge, Regular	15	22.510	7.510	56.400	2.506
Overcharge, Special	29	21.490	-7.510	56.400	2.624
Correct Charge, Regular	384	382.676	1.324	1.753	0.005
Correct Charge, Special	364	365.324	-1.324	1.753	0.005
Totals	819	819			$\chi^2 = 10.814$

e) From Table 10.12, the critical value for a 3 x 2 table is 9.210 for the 0.01 significance level. Since χ^2= 10.814 exceeds this value, the P-value is less than 0.01.

f) Reject independence at the 0.01 level and conclude that there is a relationship between the pricing and the charging.